Alternative Natural Gas Applications Workshop

Creating a Prosperous Demand Market

Alexandria, Virginia, USA
8-9 October 2014

ISBN: 978-1-5108-1765-4

Printed from e-media with permission by:

Curran Associates, Inc.
57 Morehouse Lane
Red Hook, NY 12571

Some format issues inherent in the e-media version may also appear in this print version.

Printed by Curran Associates, Inc. (2016)

For permission requests, please contact AIChE
at the address below.

AIChE
120 Wall Street, FL 23
New York, NY 10005-4020

Phone: (800) 242-4363
Fax: (203) 775-5177

www.aiche.org

Additional copies of this publication are available from:

Curran Associates, Inc.
57 Morehouse Lane
Red Hook, NY 12571 USA
Phone: 845-758-0400
Fax: 845-758-2634
Email: curran@proceedings.com
Web: www.proceedings.com

TABLE OF CONTENTS

OPENING SESSION

The Outlook for Energy: A View to 2040 1
Nick Jones
Alternative Natural Gas Applications Workshop 10
Monty Alger
U.S. Natural Gas - A Decade of Change and the Emergence of the Gas-Centric Future 23
Francis O'Sullivan

STATIONARY ENERGY GENERATION

Natural Gas Based Stationary Power Generation 32
Kevin Huang
Air Products: Natural Gas Conversion to H_2 48
Michael F. Carolan
Competitiveness of US Gas Turbine Manufacturers 49
William Day
Global CCS Institute 53
Ron Munson
Fuel Cells and Hydrogen for Greater Sustainability Using Natural Gas 58
Pinakin Patel

TRANSPORTATION ENERGY GENERATION

Natural Gas Opportunities and Applications for Sustainable Transportation 69
Reuben Sarkar

ROUNDTABLE SUMMARIES

Alternative Natural Gas Applications Workshop: Creating a Prosperous Demand Market
Transportation Roundtable Discussion 82
N/A
Chemical Synthesis 92
N/A
Stationary Power Gen Roundtable Discussion 93
N/A
Author Index

The Outlook for Energy: A View to 2040

Nick Jones, Energy Advisor

AIChE Alternative Natural Gas Applications Workshop
October 8, 2014

ExxonMobil

Energy Outlook Development

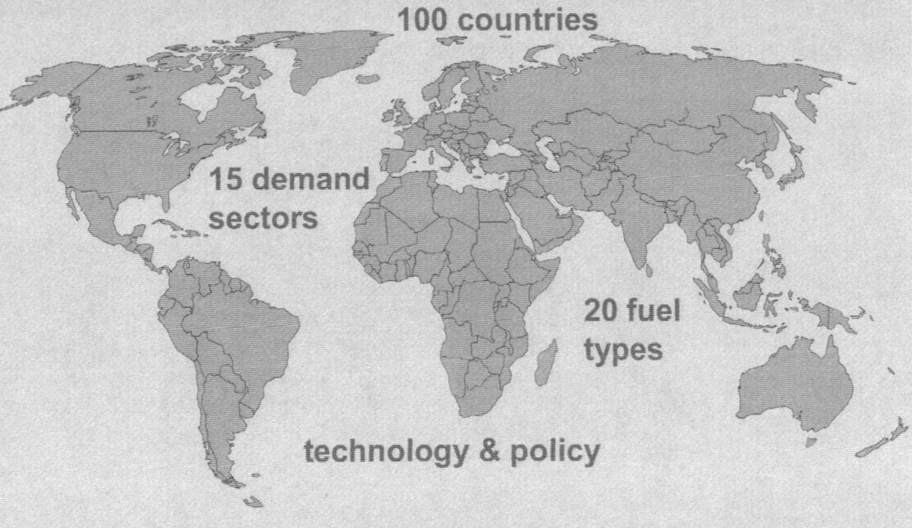

ExxonMobil 2014 Outlook for Energy

ExxonMobil

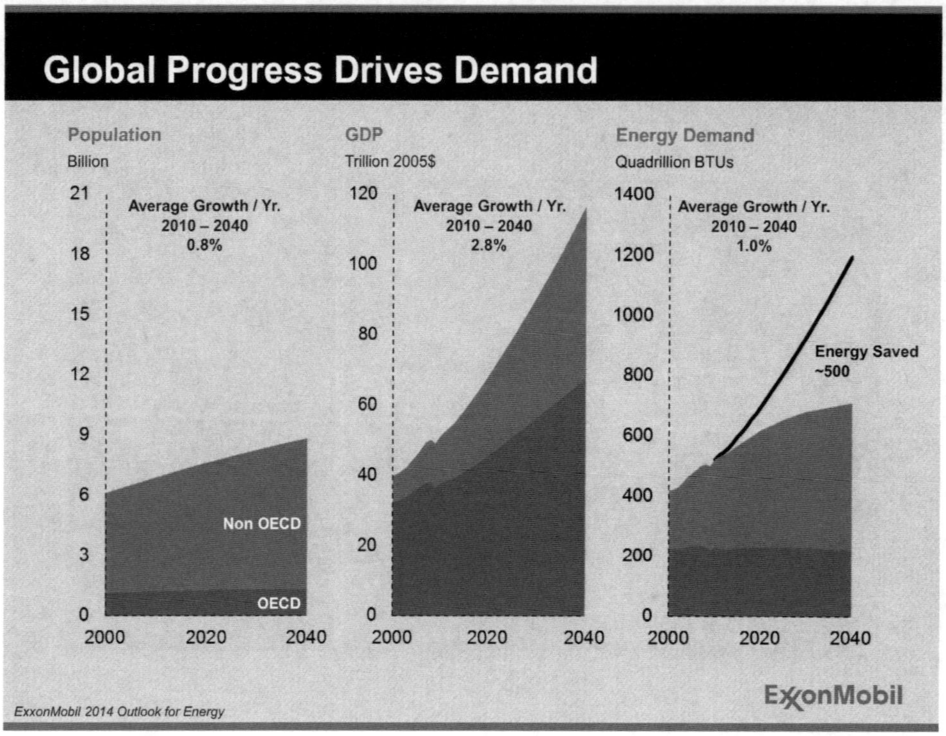
Global Progress Drives Demand
Population
Billion
Average Growth / Yr.
2010 – 2040
0.8%
Non OECD
OECD
GDP
Trillion 2005$
Average Growth / Yr.
2010 – 2040
2.8%
Energy Demand
Quadrillion BTUs
Average Growth / Yr.
2010 – 2040
1.0%
Energy Saved
~500
2000
2020
2040
ExxonMobil 2014 Outlook for Energy
ExxonMobil

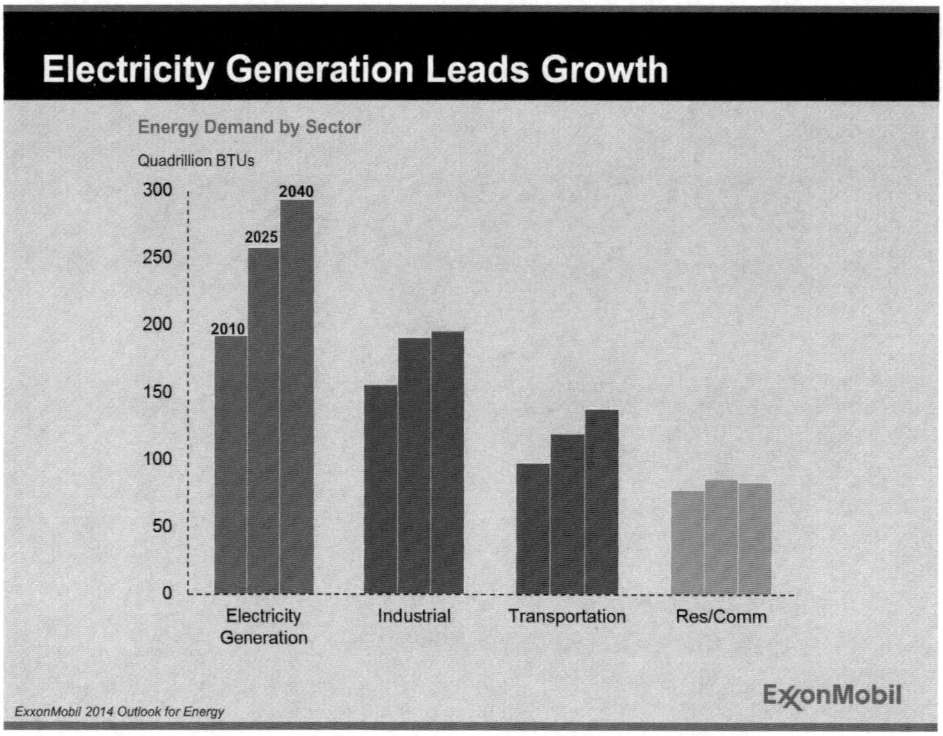
Electricity Generation Leads Growth
Energy Demand by Sector
Quadrillion BTUs
2010
2025
2040
Electricity Generation
Industrial
Transportation
Res/Comm
ExxonMobil 2014 Outlook for Energy
ExxonMobil

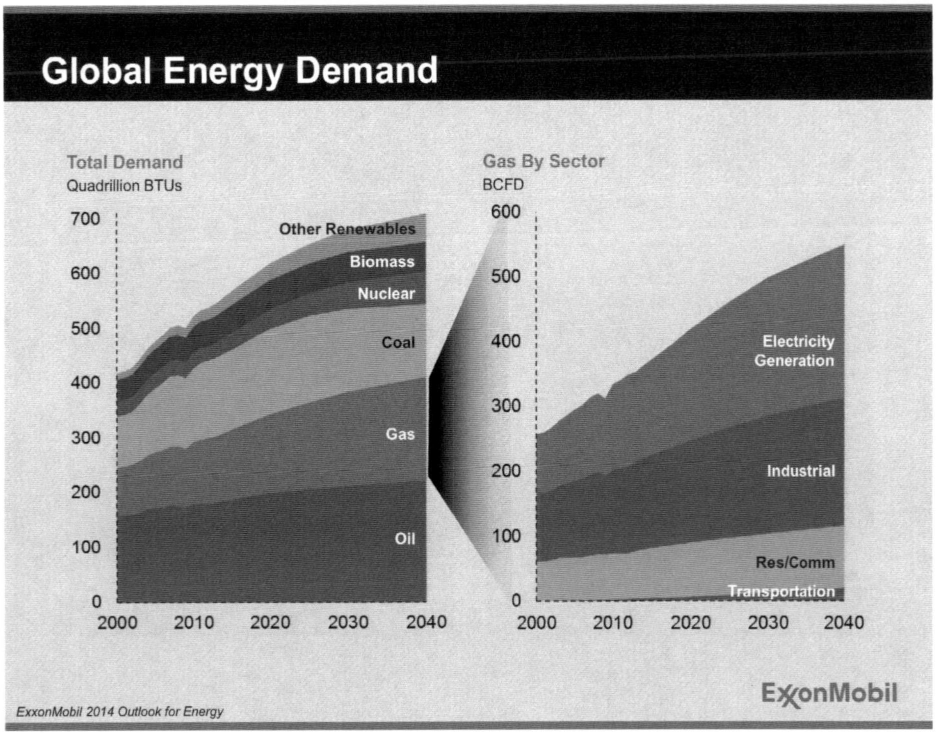
Global Energy Demand
Total Demand
Quadrillion BTUs
700
600
500
400
300
200
100
0
2000
2010
2020
2030
2040
Other Renewables
Biomass
Nuclear
Coal
Gas
Oil
Gas By Sector
BCFD
600
500
400
300
200
100
0
2000
2010
2020
2030
2040
Electricity Generation
Industrial
Res/Comm
Transportation
ExxonMobil 2014 Outlook for Energy
ExxonMobil

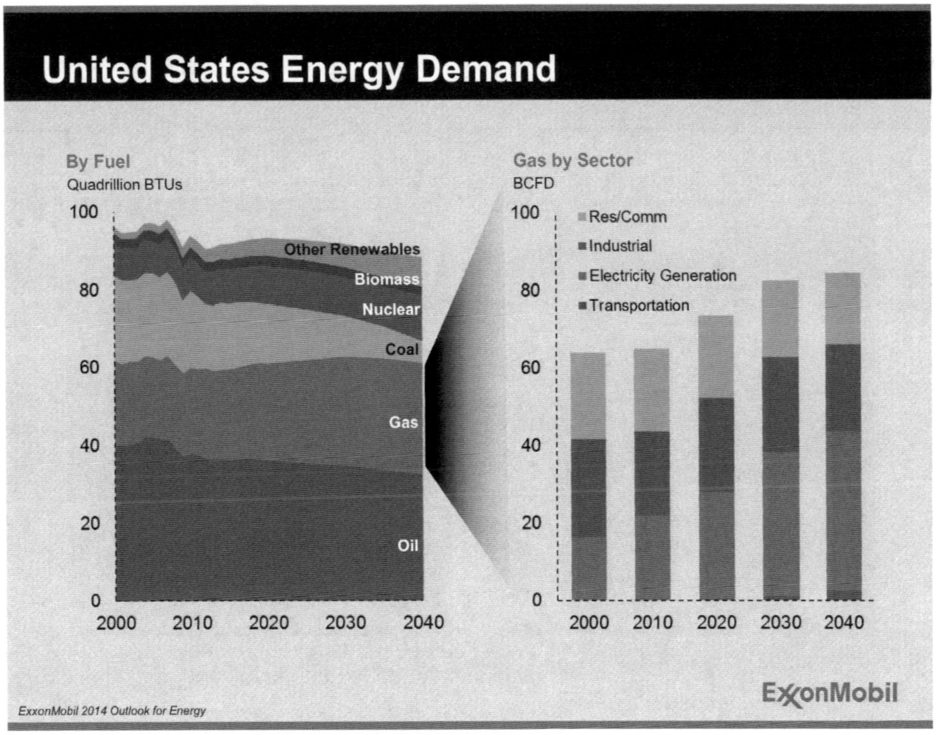
United States Energy Demand
By Fuel
Quadrillion BTUs
100
80
60
40
20
0
2000
2010
2020
2030
2040
Other Renewables
Biomass
Nuclear
Coal
Gas
Oil
Gas by Sector
BCFD
100
80
60
40
20
0
Res/Comm
Industrial
Electricity Generation
Transportation
2000
2010
2020
2030
2040
ExxonMobil 2014 Outlook for Energy
ExxonMobil

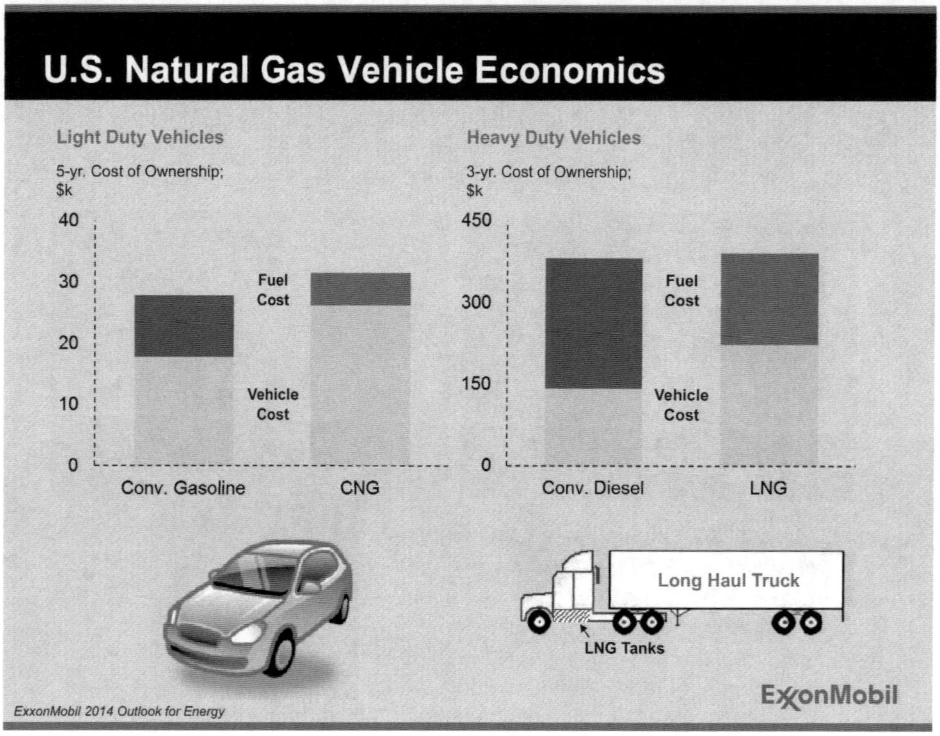
U.S. Natural Gas Vehicle Economics
Light Duty Vehicles
5-yr. Cost of Ownership; $k
40
30
20
10
0
Fuel Cost
Vehicle Cost
Conv. Gasoline
CNG
Heavy Duty Vehicles
3-yr. Cost of Ownership; $k
450
300
150
0
Fuel Cost
Vehicle Cost
Conv. Diesel
LNG
Long Haul Truck
LNG Tanks
ExxonMobil 2014 Outlook for Energy
ExxonMobil

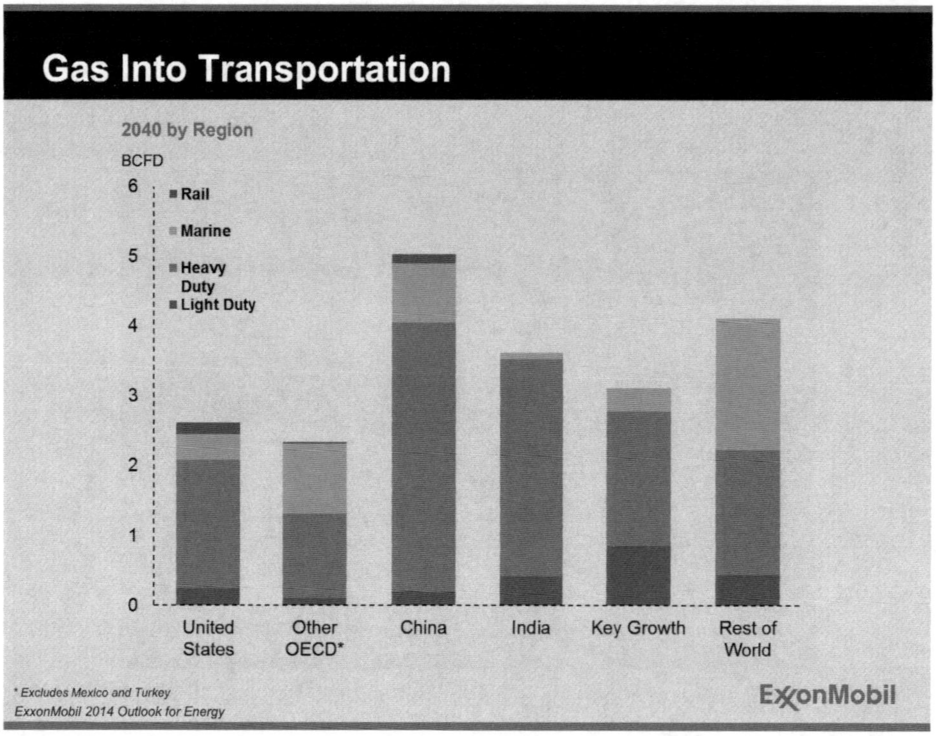
Gas Into Transportation
2040 by Region
BCFD
6
5
4
3
2
1
0
Rail
Marine
Heavy Duty
Light Duty
United States
Other OECD*
China
India
Key Growth
Rest of World
* Excludes Mexico and Turkey
ExxonMobil 2014 Outlook for Energy
ExxonMobil

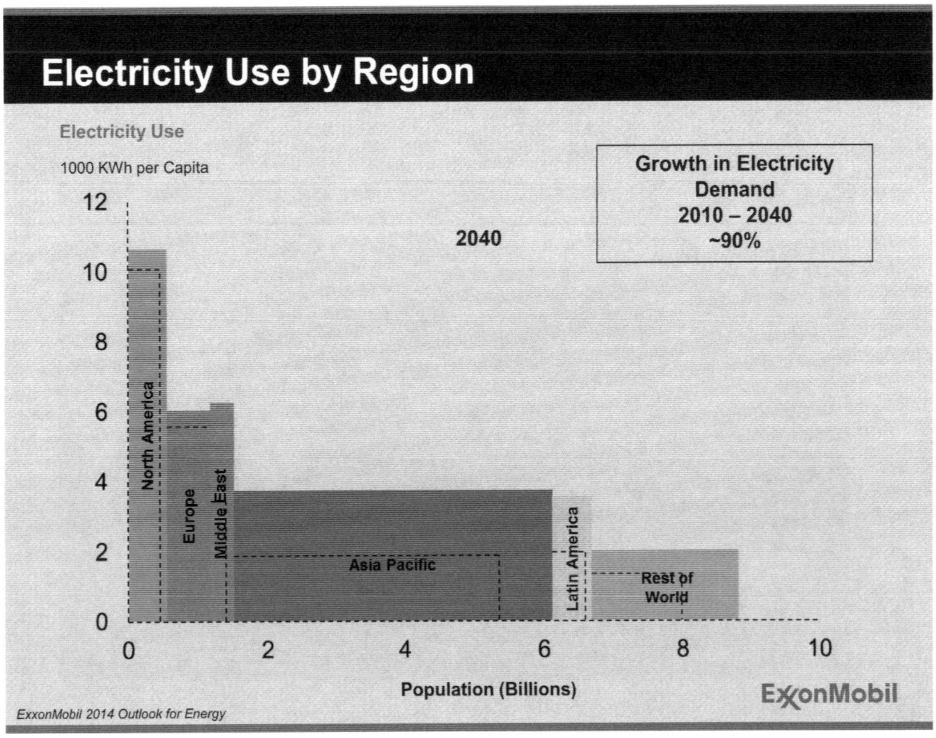
Electricity Use by Region
Electricity Use
1000 KWh per Capita
12
10
8
6
4
2
0
2040
Growth in Electricity Demand
2010 – 2040
~90%
North America
Europe
Middle East
Asia Pacific
Latin America
Rest of World
0
2
4
6
8
10
Population (Billions)
ExxonMobil
ExxonMobil 2014 Outlook for Energy

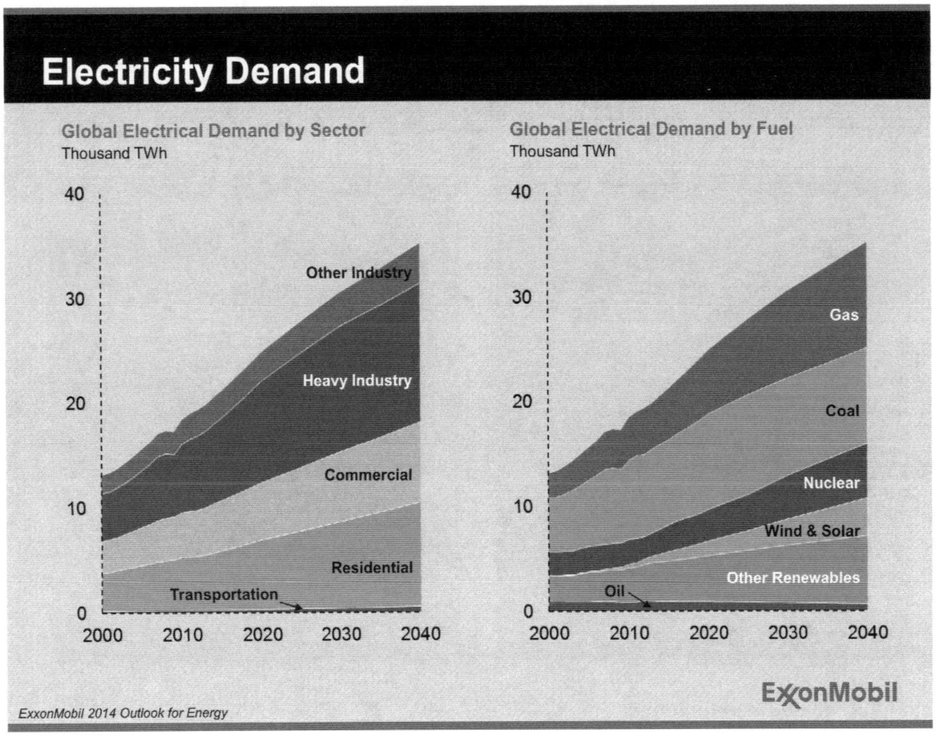
Electricity Demand
Global Electrical Demand by Sector
Thousand TWh
40
30
20
10
0
Other Industry
Heavy Industry
Commercial
Residential
Transportation
2000
2010
2020
2030
2040
Global Electrical Demand by Fuel
Thousand TWh
40
30
20
10
0
Gas
Coal
Nuclear
Wind & Solar
Other Renewables
Oil
2000
2010
2020
2030
2040
ExxonMobil
ExxonMobil 2014 Outlook for Energy

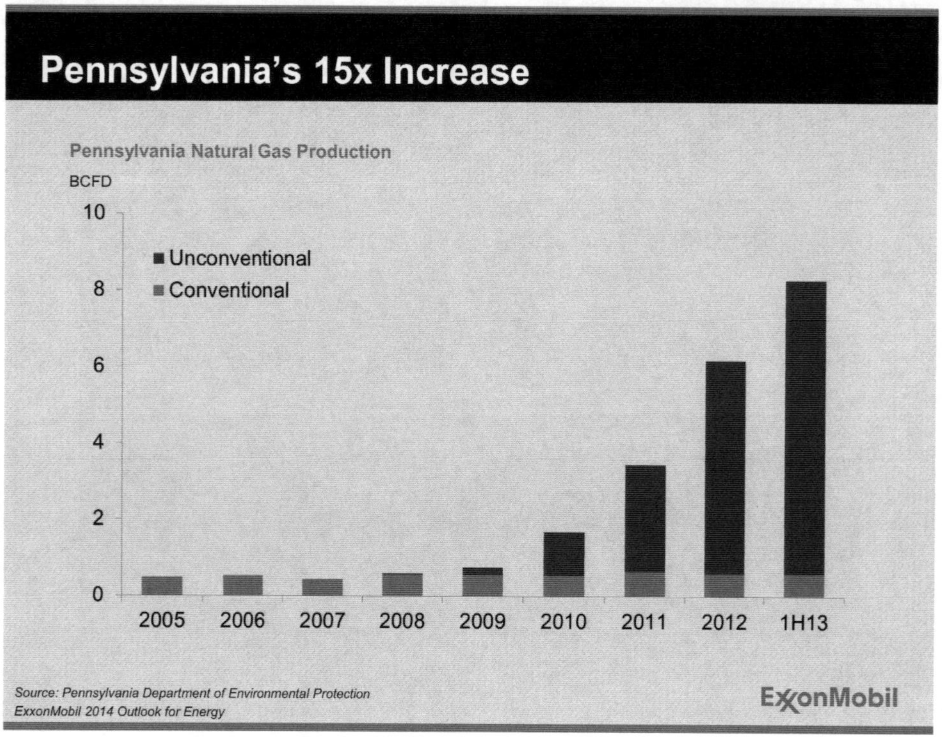
Pennsylvania's 15x Increase
Pennsylvania Natural Gas Production
BCFD
10
8
6
4
2
0
Unconventional
Conventional
2005
2006
2007
2008
2009
2010
2011
2012
1H13
Source: Pennsylvania Department of Environmental Protection
ExxonMobil 2014 Outlook for Energy
ExxonMobil

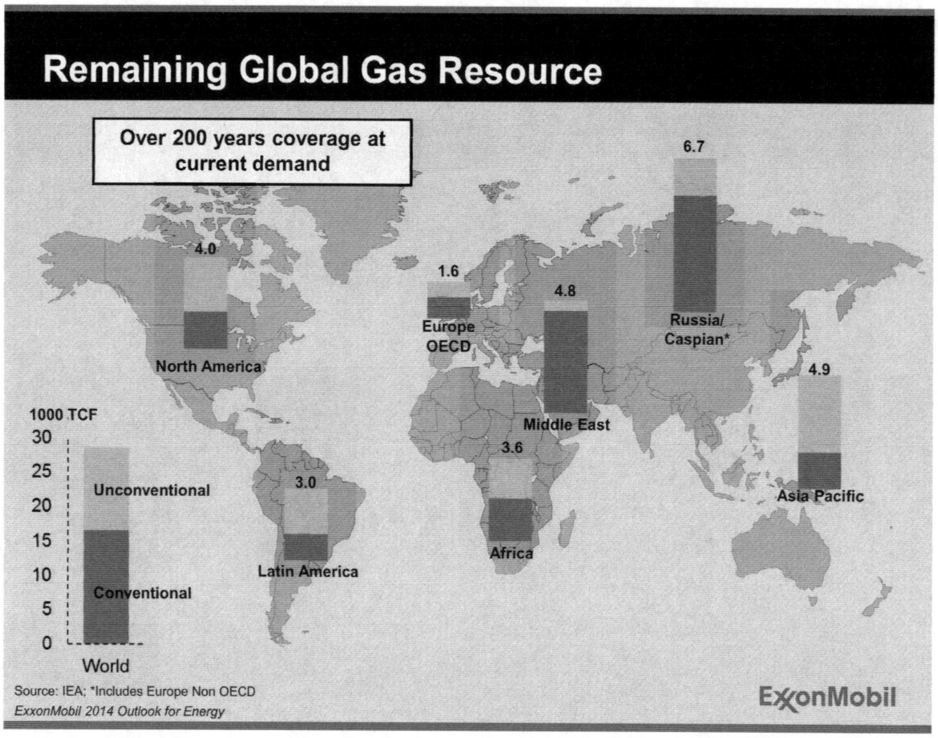
Remaining Global Gas Resource
Over 200 years coverage at current demand
4.0
North America
1.6
Europe OECD
4.8
Middle East
6.7
Russia/ Caspian*
4.9
Asia Pacific
3.6
Africa
3.0
Latin America
1000 TCF
30
25
20
15
10
5
0
Unconventional
Conventional
World
Source: IEA; *Includes Europe Non OECD
ExxonMobil 2014 Outlook for Energy
ExxonMobil

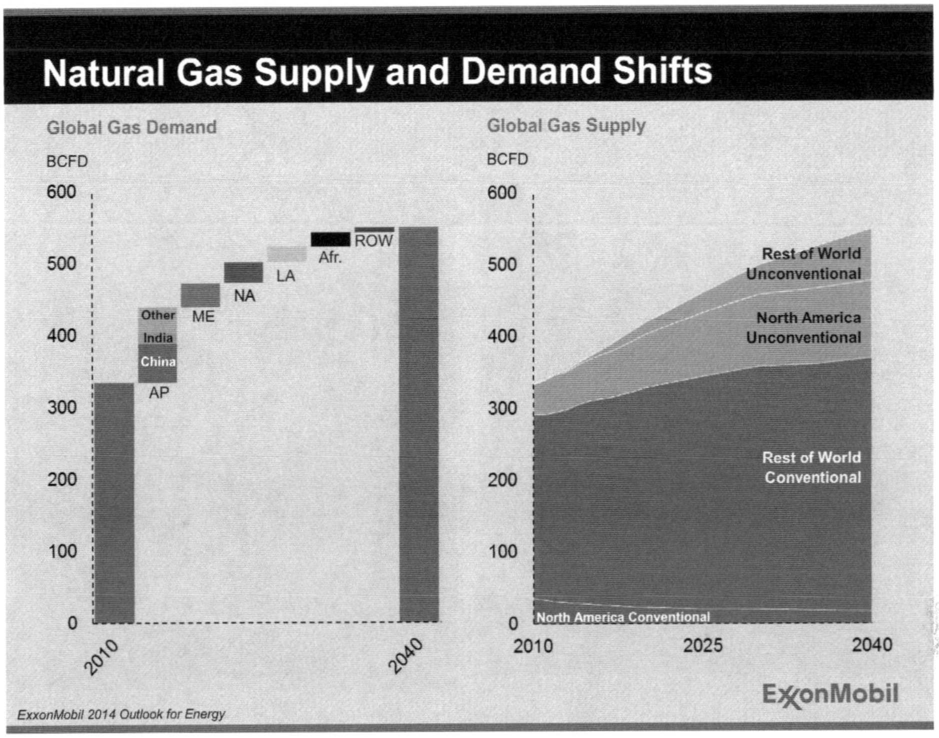

Natural Gas Supply and Demand Shifts
Global Gas Demand
BCFD
600
500
400
300
200
100
0
2010
2040
AP
China
India
Other
ME
NA
LA
Afr.
ROW
Global Gas Supply
BCFD
600
500
400
300
200
100
0
2010
2025
2040
Rest of World Unconventional
North America Unconventional
Rest of World Conventional
North America Conventional
ExxonMobil 2014 Outlook for Energy
ExxonMobil

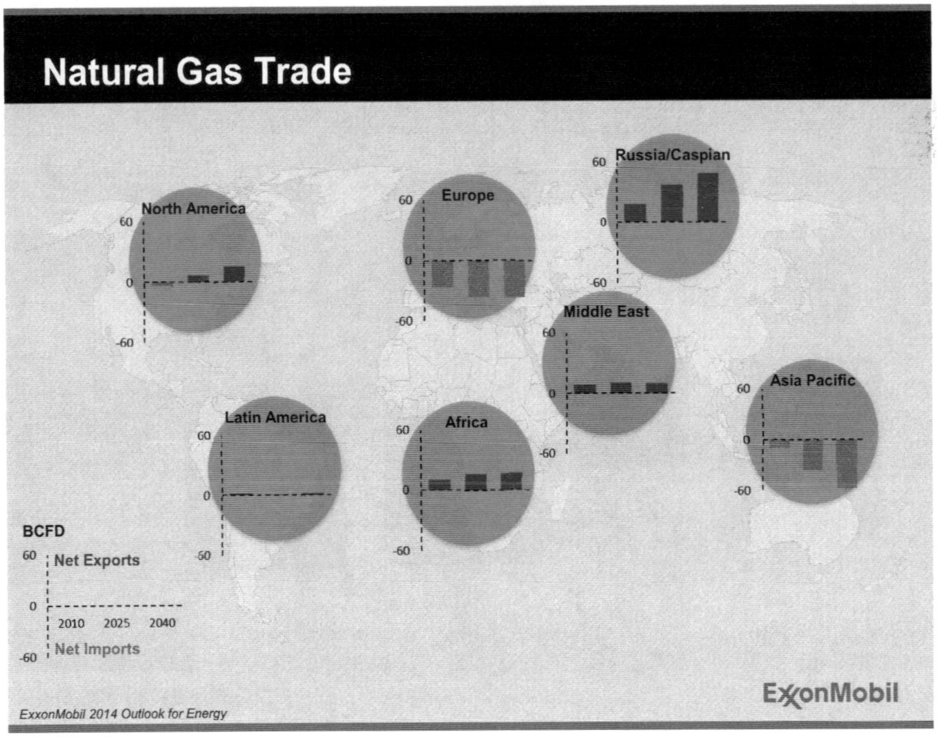

Natural Gas Trade
North America
Europe
Russia/Caspian
Middle East
Asia Pacific
Latin America
Africa
60
0
-60
BCFD
60 Net Exports
0
2010 2025 2040
-60 Net Imports
ExxonMobil 2014 Outlook for Energy
ExxonMobil

Regional Energy Trends Evolve

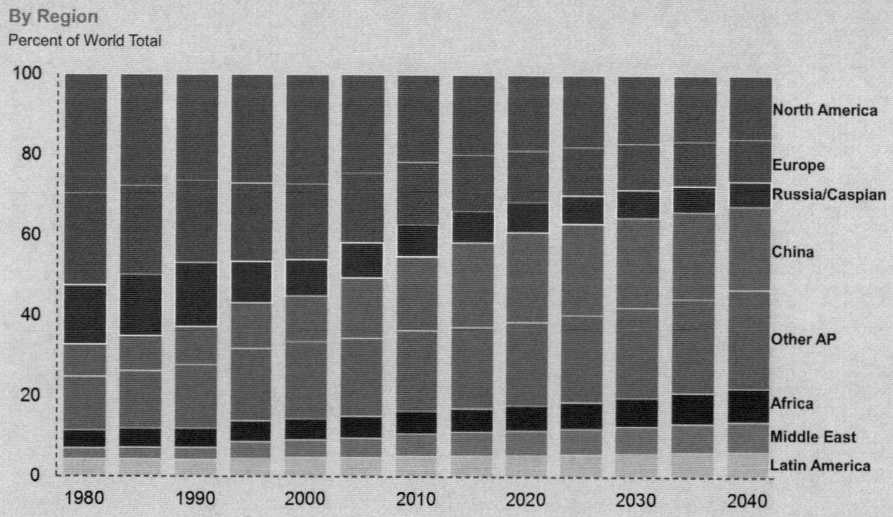

ExxonMobil 2014 Outlook for Energy

ExxonMobil

Conclusions

ExxonMobil 2014 Outlook for Energy

ExxonMobil

ExxonMobil

ExxonMobil

PENNSTATE

Alternative Natural Gas Applications Workshop
Creating a Prosperous Demand Market

10-8-2014

Monty Alger
Penn State
Director, Institute for Natural Gas Research

PENNSTATE

Future Goals for Energy

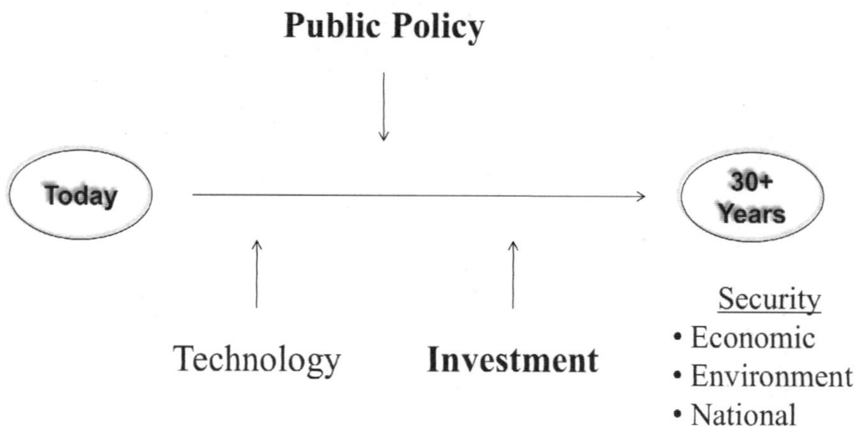

Energy and Economy

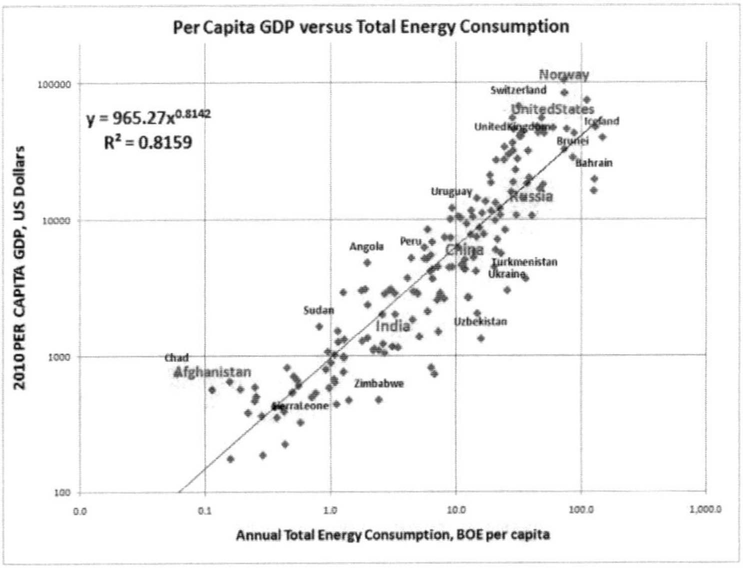

3

Unconventional Oil and Gas – Everything has Changed

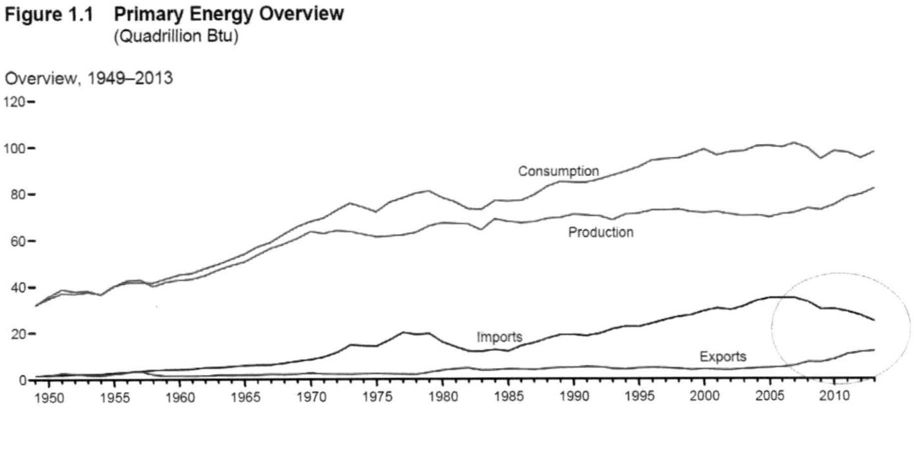

EIA Monthly Energy Review 9/2014
http://www.eia.gov/totalenergy/data/monthly/pdf/mer.pdf

4

Shale Gas Production

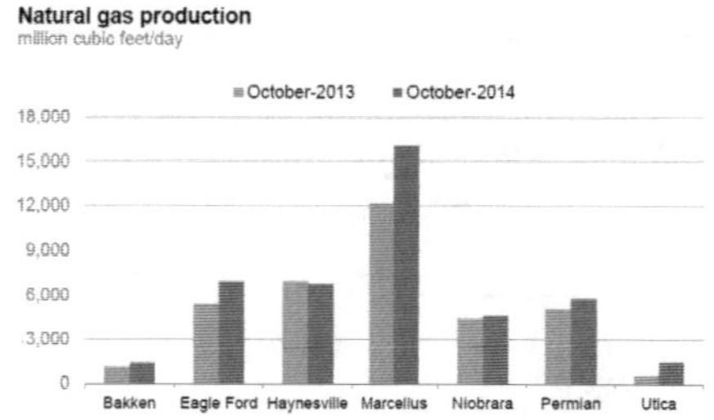

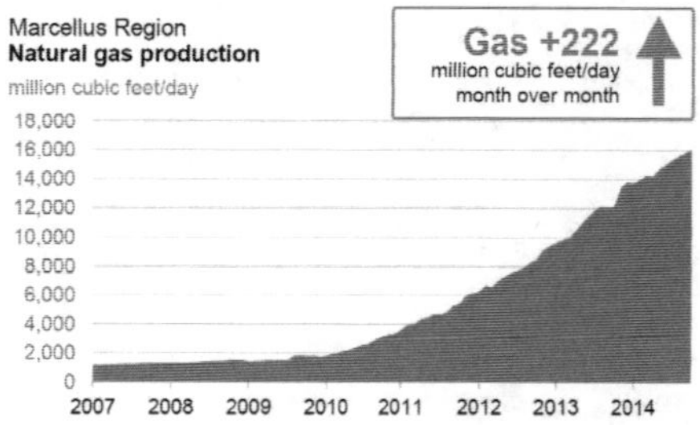

- US Shale Gas Production
- Production quadrupled 2 years

Ref: http://www.eia.gov/petroleum/drilling/pdf/marcellus.pdf

5

Estimated US Energy Use 2013

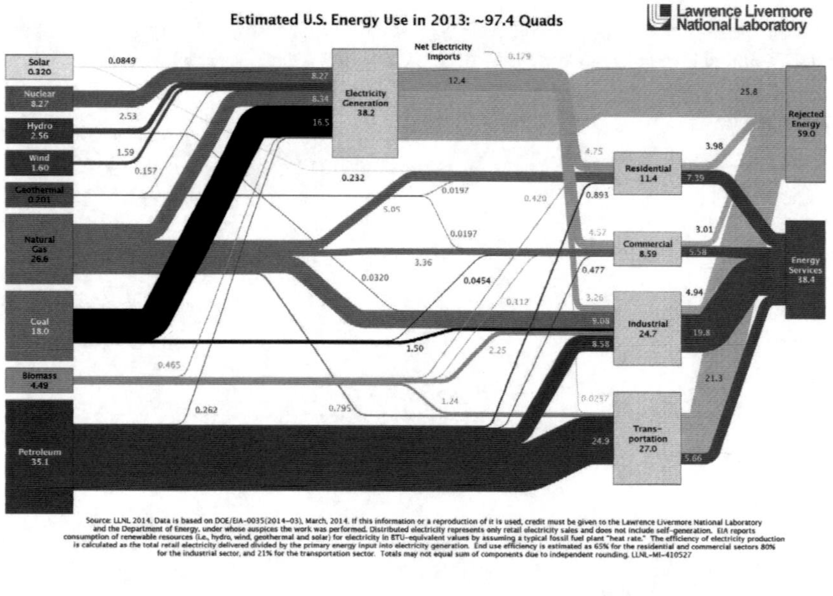

6

PENNSTATE

Estimated US Carbon Emissions 2013

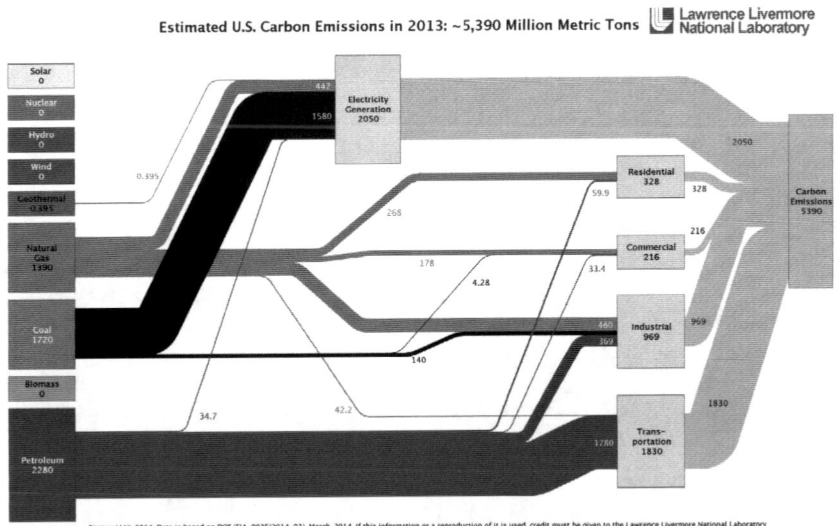

Source: LLNL 2014. Data is based on DOE/EIA-0035(2014-03), March, 2014. If this information or a reproduction of it is used, credit must be given to the Lawrence Livermore National Laboratory and the Department of Energy, under whose auspices the work was performed. Carbon emissions are attributed to their physical source, and are not allocated to end use for electricity consumption in the residential, commercial, industrial and transportation sectors. Petroleum consumption in the electric power sector includes the non-renewable portion of municipal solid waste. Combusition of biologically derived fuels is assumed to have zero net carbon emissions - the lifecycle emissions associated with producing biofuels are included in commercial and industrial emissions. Totals may not equal sum of components due to independent rounding errors. LLNL-MI-410527

7

PENNSTATE

US Energy Sources … Fossil Fuels ~82%

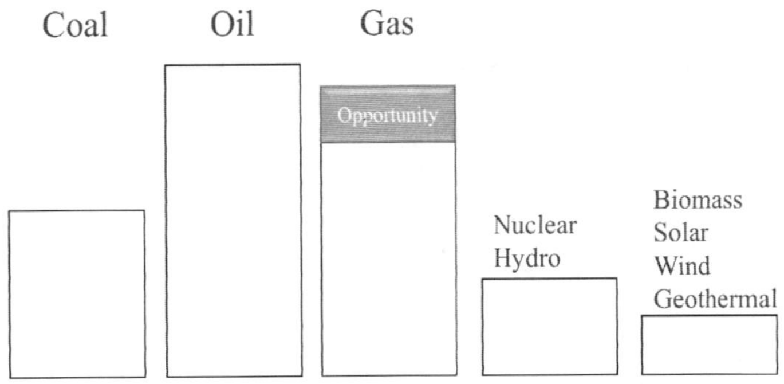

	Coal	Oil	Gas	Nuclear Hydro	Biomass Solar Wind Geothermal
% US	19%	36%	27%	11%	7%
$/ MM BTU*	2.4	12	4.3		
CO_2/BTU	100%	70%	50%	~0%	~0%

*Cost of fossil fuels at electricity generating plant 2013, U.S. Energy Information Administration / Monthly Energy Review September 2014 page 129

PENNSTATE

Public Policy … New Boundary Conditions

Air Pollution and the Clean Air Act

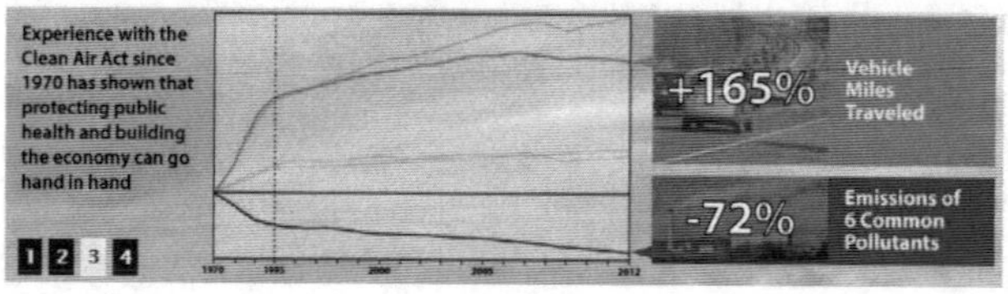

Clean Water Act

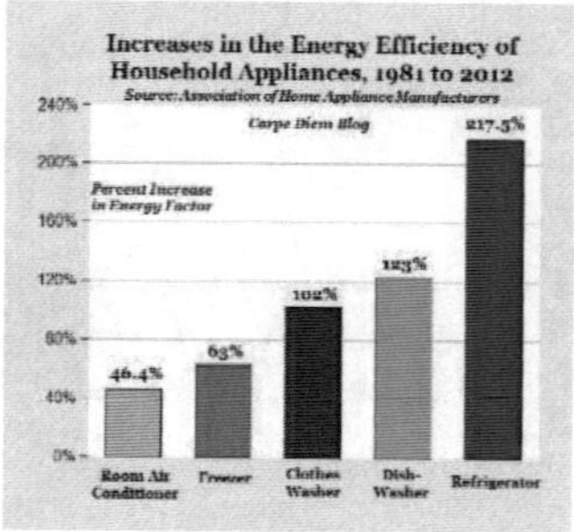

9

Which is More "Sustainable" ?

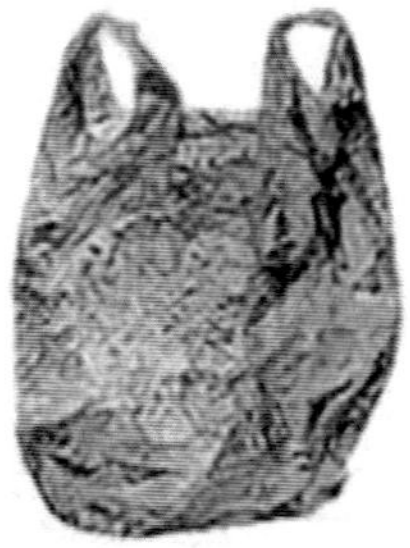

Paper Bag (Recycled Content)

Plastic Bag

Footprint for Paper and Plastic Bags

Impact Summary of Various Bag Types (Carrying Capacity Equivalent to 1000 Paper Bags)	**Paper (30% Recycled Fiber)**	**Polyethylene**
Total Energy Usage (MJ)	2,622	763
Fossil Fuel Use (kg)	23.2	14.9
Municipal Solid Waste (kg)	33.9	7.0
Greenhouse Gas Emissions (CO_2 Equiv. Tons)	0.08	0.04
Fresh Water Usage (Gal)	1,004	58
Landfill Half-Life	Short	?

Sources:

- "**Bag fact sheet**", American Chemistry Council
- " **Life Cycle Assessment for Three Types of Grocery Bags - Recyclable Plastic; Compostable, Biodegradable Plastic; and Recycled, Recyclable Paper** ", Boustead Consulting & Associates Ltd

Life Cycle Assessment

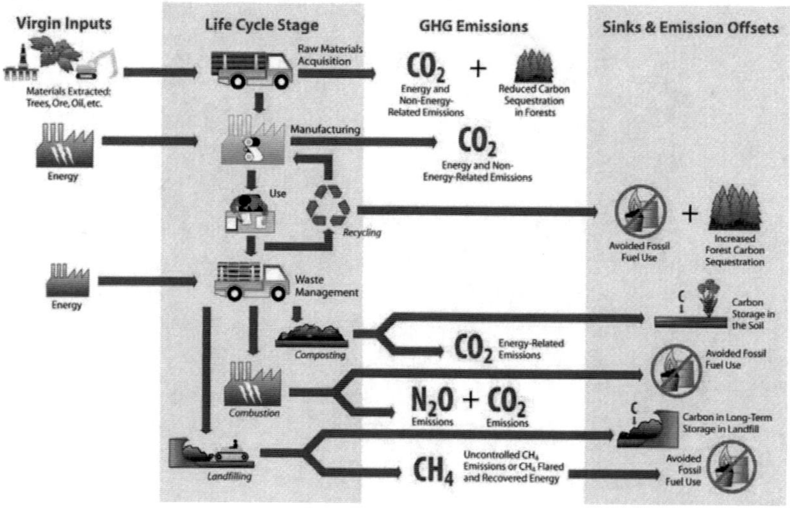

Source: EPA, http://www.epa.gov/statelocalclimate/images/flow_chart_life_cycle_big.jpg

Summary – Push to the "A" Box

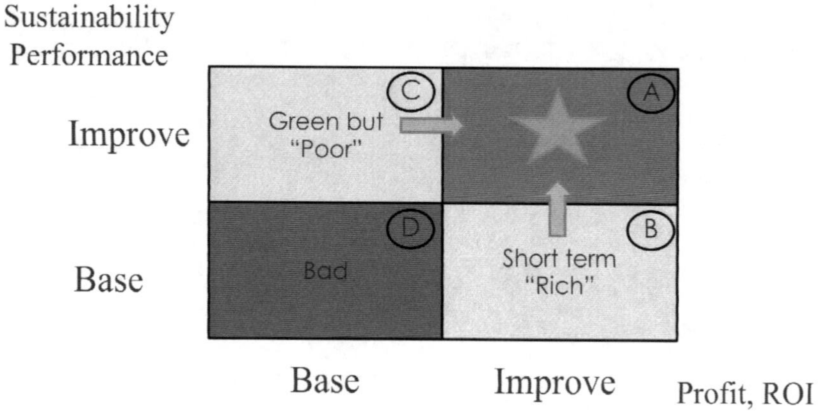

- "Shale Gas - new major investments, environmental improvement
- Sustainability – an evolving change to boundary conditions for solution
- Innovation – push for the (A) box ; technology (solution) <-> Need (market)

PENNSTATE

Chemicals – Shell Cracker

PENNSYLVANIA Energy. Environment. Economy.

REGULATION ECONOMY INFLUENCE IMPACT

Shell actively exploring Western Pa. cracker project

NOVEMBER 19, 2013 | 5:30 PM
BY KATIE COLANERI

Royal Dutch Shell's decision about whether to build a multibillion dollar ethane cracker in Beaver County could still be several years off, but the Associated Press has found that the company is actively taking steps to explore the site.

More from the Associated Press:

PENNSYLVANIA Energy. Environment. Economy.

REGULATION ECONOMY INFLUENCE IMPACT

Shell's proposed ethane cracker could be a big emitter

JUNE 23, 2014 | 12:45 PM
BY KATIE COLANERI

The Pittsburgh Tribune-Review reports Shell's proposed ethane cracker in western Pennsylvania could be a significant new source of air pollution.

The company has not yet committed to building the multi-billion dollar petrochemical plant, but has applied to the Department of Environmental Protection for an air quality permit for the site in Monaca, Beaver County.

Shell construction manager Ken Conly talks about the ethane cracker project at a public meeting.

Potential New Chemicals Investment

Local / Global

Ref: http://stateimpact.npr.org/pennsylvania/tag/ethane-cracker/

15

PENNSTATE

Life Cycle Assessment – Horizontal not Vertical

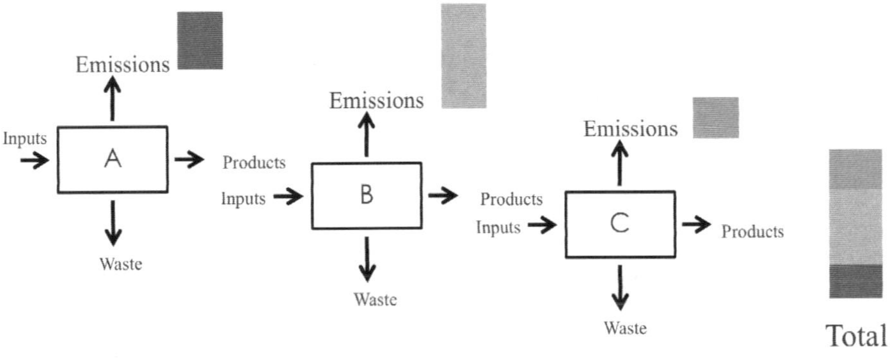

PENNSTATE

Delta Jet Fuel Sourcing

Delta Airlines' Trainer, Pennsylvania Refinery

- Invest ~150 MM$
- Savings ~ 300 MM$
- System savings

TRANSPORTATION

More: Airlines Delta Air lines Fuel Costs Transportation

How Delta Bought A Refinery And Wound Up Saving Its Rivals A Ton Of Cash

BENJAMIN ZHANG

Delta sources Bakken crude for Pa. refinery

Deal accounts for one third of refinery's capacity.

By Daniel J. Graeber | July 21, 2014 at 10:19 AM | 0 Comments (Leave a comment)

Ref: http://www.businessinsider.com/delta-airlines-fuel-prices-2014-8
http://www.upi.com/Business_News/Energy-Resources/2014/07/21/Delta-sources-Bakken-crude-for-Pa-refinery/5281405951764/

PENNSTATE

Transportation Fuel – Natural Gas

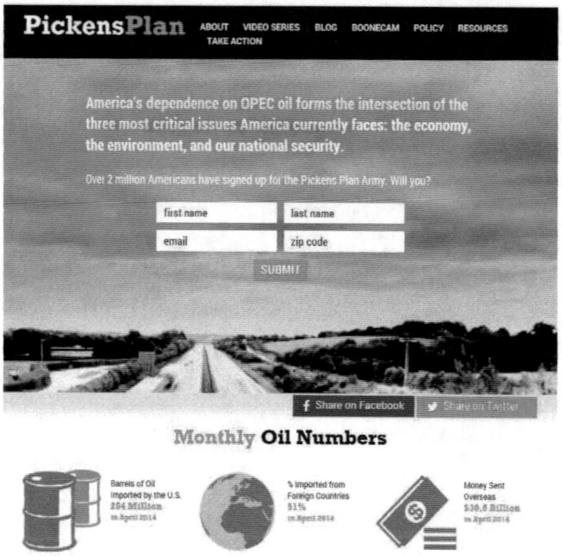

Ref: http://www.pickensplan.com

18

Transportation Example

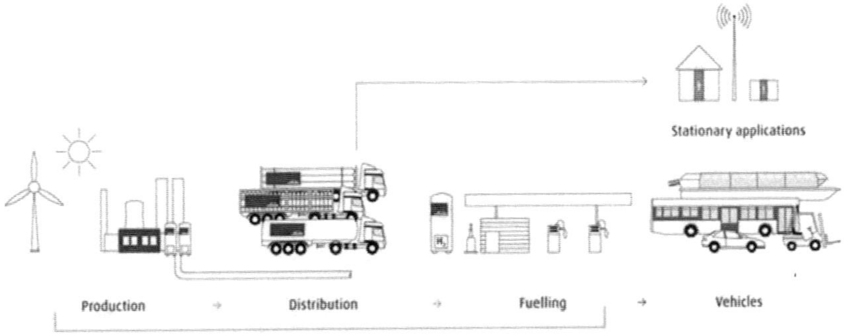

Technology | Network

Natural Gas ⇨

- Compressed Natural Gas (CNG)
- Liquefied Natural Gas (LNG)
- Dimethyl Ether (DME)
- Hydrogen

- Gas to Liquids

- Production, distribution
- Fueling stations
- Vehicles: car, rail, marine …

Ref: http://www.linde-gas.com/en/innovations/hydrogen_energy/index.html

Information Technology Example

Technology | Network

⇧

Infrastructure, standards

Natural Gas Growth

Technology
- Projects that have direct financial return
- Power generation
- Novel supply chain investments
- Residential and industrial conversion

Network
- Need coordinated supply chain investment
- Transportation fuels - infrastructure
- H2, CNG, LNG, DME
- GTL

Risk, Unknown
- Long term price of natural gas
- Sustainability – regulation, emissions, water
- High capital – risk

21

System Level – Sustainability and Profit

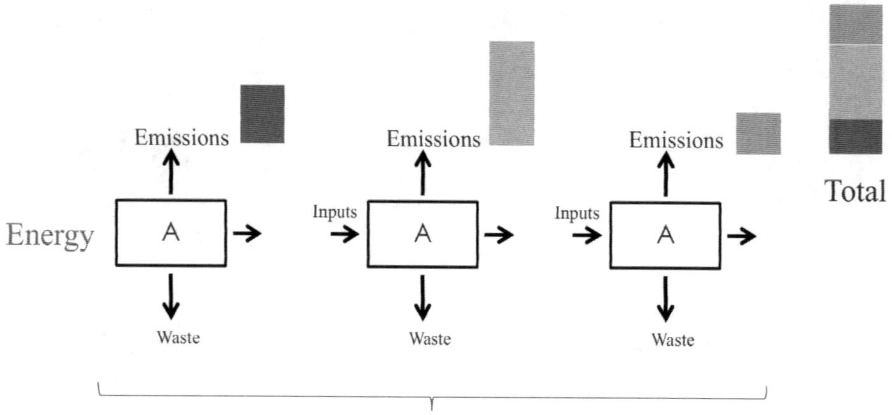

- System Level Benefits
- Operating Expense
- Capital Expense

$’s

Future Goals for Energy

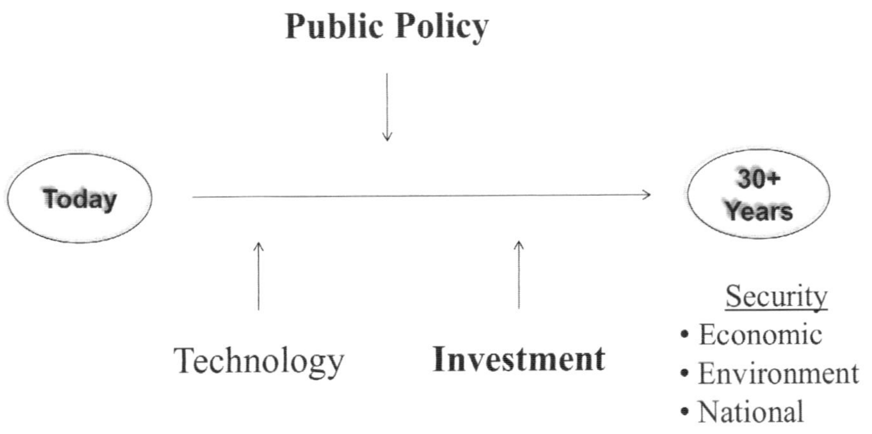

- Technology – create new opportunity
- System level thinking
- Profitability and Sustainability

Penn State – Institute for Natural Gas Research

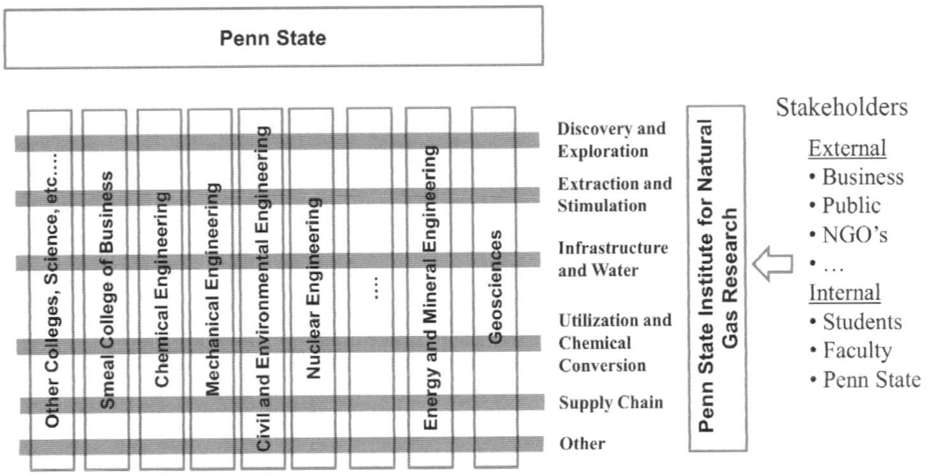

Access Penn State Capabilities Across the University
Connect with External Markets and Stakeholders

24

PENNSTATE

Thank You!

25

AIChE – Alternative Natural Gas Applications Workshop

U.S. Natural Gas – A Decade of Change and the Emergence of the Gas-Centric Future

Francis O' Sullivan

October 8th, 2014

1

Natural gas – The new "unconventional" paradigm

2

The past decade has been a period of huge change for natural gas in the United States – Perspectives on supply and price have been fundamentally altered and a much more gas-centric future is being envisaged by many

Comparison of spot natural gas price with historical oil-to-gas ratios
$/MMBtu of gas

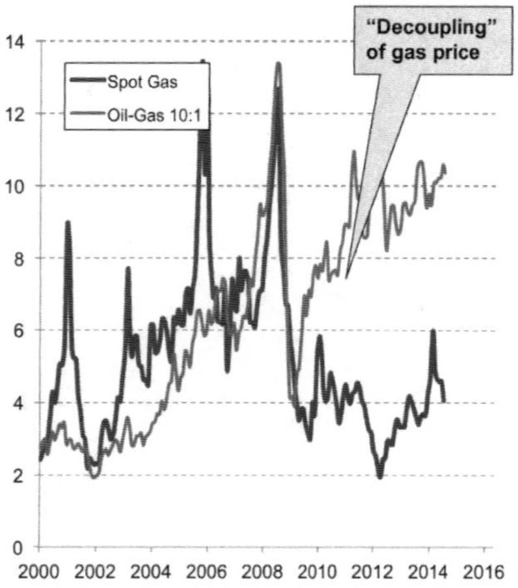

Illustration of gas production growth from the main U.S. shale plays since 2005
Bcf of gas per day

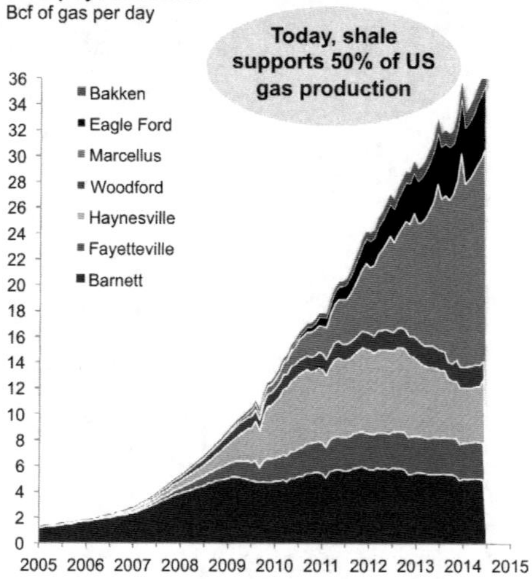

Source: F. O'Sullivan, United States Energy Information Administration, HPDI Production Database

3

However, the fact that shale gas production is still in its infancy means that large uncertainties surround resource estimates – As more well data becomes available the uncertainty envelope will likely narrow

Illustration of growth in US natural gas proved reserve and resource estimates from '90 to '10
Tcf of gas

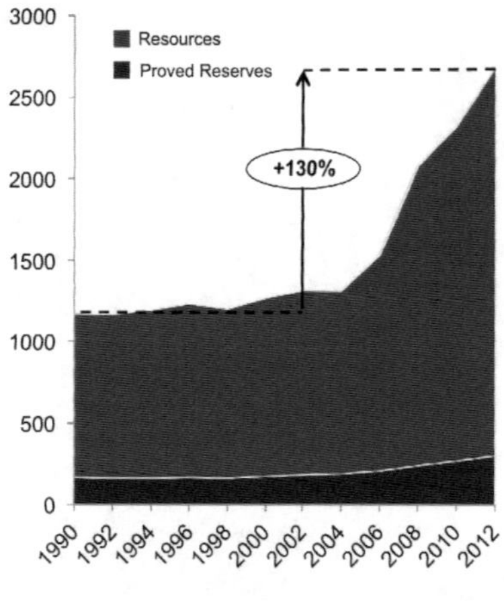

Breakdown of the PGC 2012 shale gas resource estimates by major U.S. shale play*
Tcf of Gas

	Min	Most Likely	Max
Fort Worth Basin: Barnett Shale	11	48	83
Arkoma Basin: Fayetteville & Woodford	75	104	137
E. TX & LA Basin: Haynesville & Bossier	76	149	293
TX Gulf Coast Basin: Eagle Ford & Pearsall	29	59	105
Appalachian Basin: Marcellus, Ohio & Utica	220	563	1242
Uinta Basin: Mancos & Manning Canyon	37	60	129
Other Basins:	34	90	234
Total Mean Estimate:	482**	**1073**	2223**

1. EIA 2010 assessment based on 2008 PGC assessment with updated estimates of technically recoverable shale gas volumes
Source: F. O'Sullivan, NPC data, PGC data, EIA data

4

The shale formations supporting the recent production growth are essentially source rocks – The physics underlying production from a shale setting is very different to that of conventional gas reservoir

Darcy's Law – A fundamental relationship in petroleum engineering	$q = -\frac{k}{\mu}\nabla P$	q = Fluid flux k = Permeability μ = Viscosity ∇P = Pressure gradient

Fine sandstone

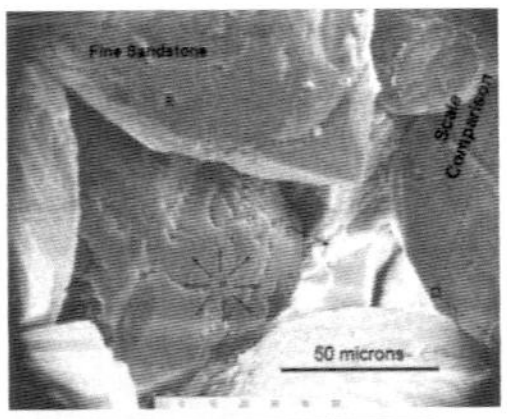

Shale

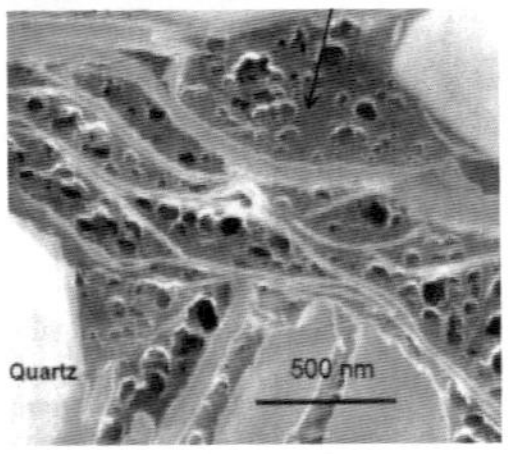

Pore throat size spectrum

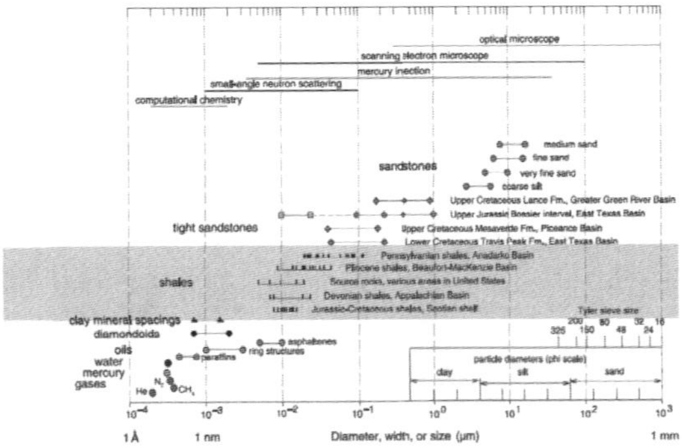

$k_{Sandstone}$ 10-100's mD k_{Shale} 10-100's nD

Source: P. Nelson, Pore-throat sizes in sandstones, tight sandstones, and shales, AAPG Bulletin, v. 93, no. 3 (March 2009), pp. 329–340

5

The combination of horizontal drilling and hydraulic fracturing is required to achieve an acceptable flow rate from a shale formation – Stimulation is not new but the scale of today's treatments are an order of magnitude larger

A fracturing stage pumping and pressure profile

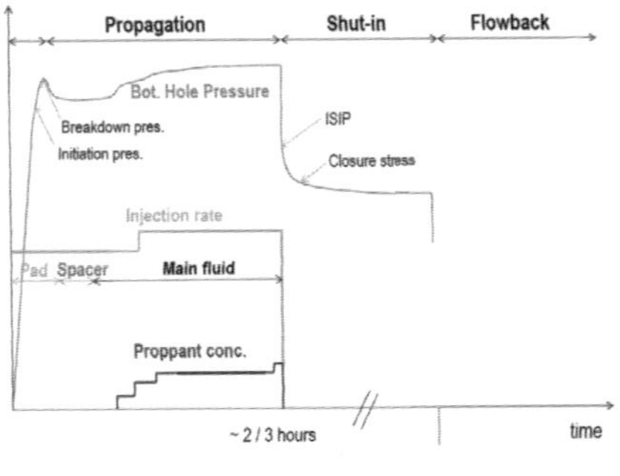

A contemporary unconventional well layout

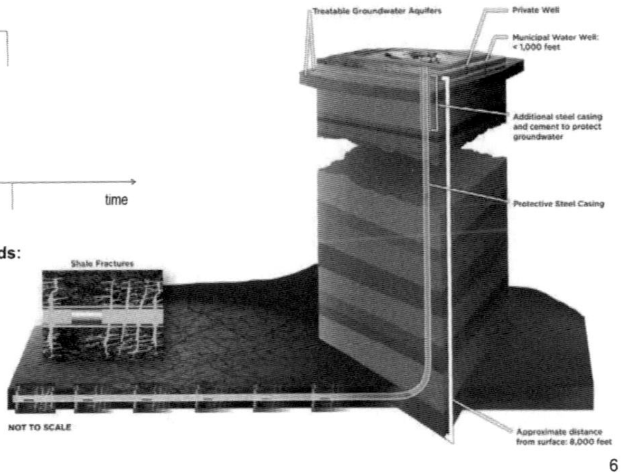

Hydraulic fracturing a single well demands:

- Horse power – 20-30,000 HP
- Pressures – 4-8,000 psi
- Water – 4-6 M gallons
- Sand – 1-2,000 Tons

Source: F. O'Sullivan, Brice Lecampion

6

The rise of U.S. shale oil and gas production has led to a large increase in the number of hydraulic fracture treatments – With this has come an increased focus on the array of complex environmental issues associated with the process

Horizontal wells completed in major U.S. shale plays
Annual well completions

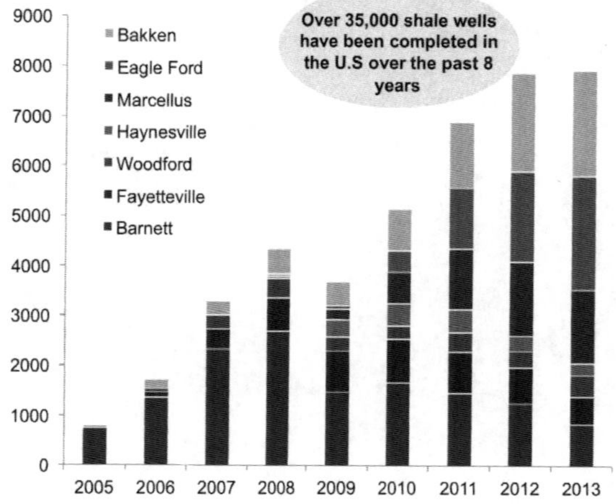

Some of the environmental issues associated with hydraulic fracturing

- **Water impacts**
  - Ground water and surface water contamination
  - Very large and impulsive demand on limited local resources
- **Air impacts**
  - Fugitive methane leakage
  - VOC emissions and other local air quality impacts
- **Community impacts**
  - Heavy traffic and surface disturbance
  - Ecosystem fragmentation
  - Induced seismicity

Source: F. O'Sullivan, HPDI Production Database

7

Understanding the shale resource – Productivity and economics

8

Reviewing early-life well performance across the major shale plays reveals some interesting features – Well productivity distributions tend to be broad and all display positive skew.

Distribution of absolute peak month well productivity[1]
Bakken H wells drilled between 2010 and 2012

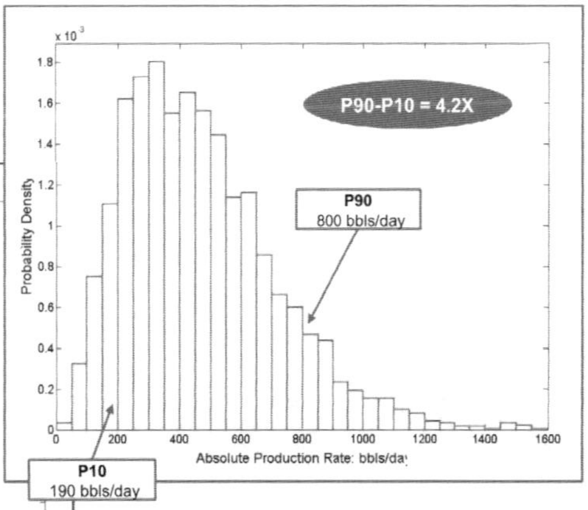

Distribution of absolute peak month well productivity[1]
Barnett H wells drilled between 2005 and 2012

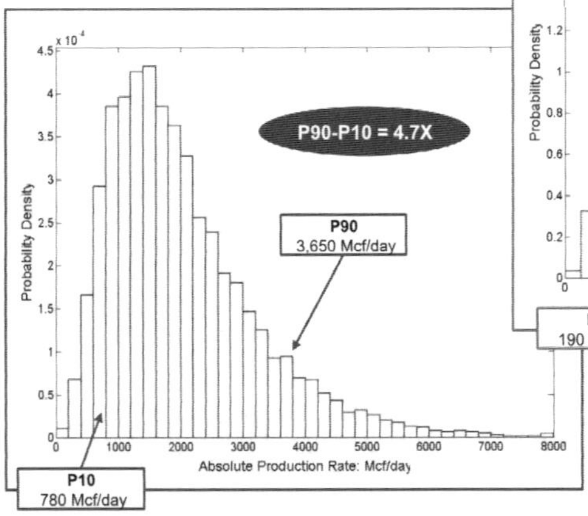

Identifying the drivers of performance variability is exceptionally difficult

- Impact of geological variation
- Impact of well completion design
- Temporal impact of a creaming process
- Etc.

1. Peak month production rate reported in units of Mcf/day and bbls/day
Source: F. O'Sullivan, HPDI database

9

Shale has provided the U.S. with an abundance of moderate cost gas, with 500 Tcf or more available at or below $6.00/MMBtu – Although often suggested as such, shale gas is not cheap

Aggregate United States natural gas supply curve
$/MMBtu breakeven gas price*

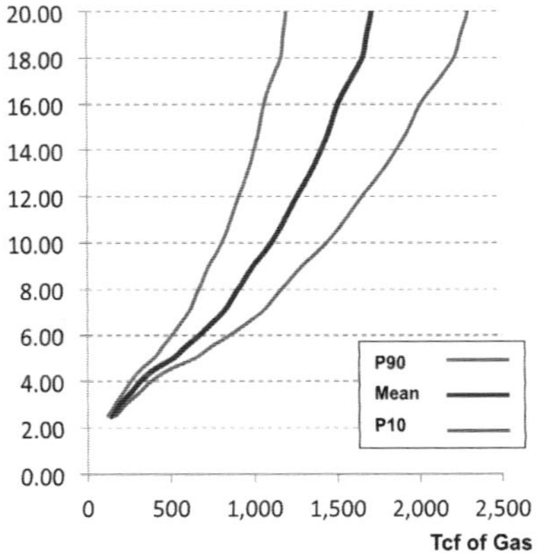

Breakdown of United States natural gas supply curves by resource type
$/MMBtu breakeven gas price*

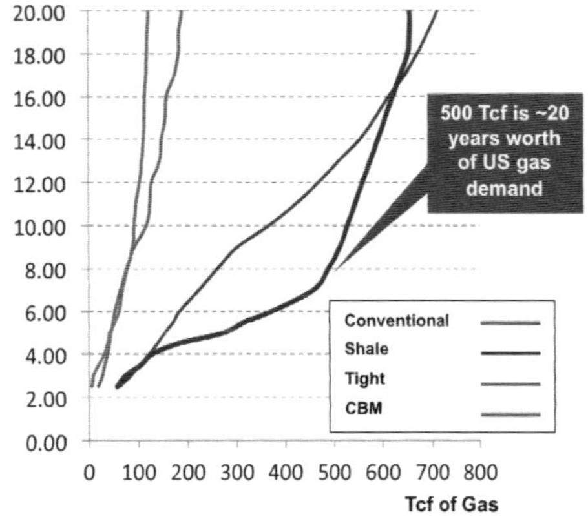

* Cost curves calculated using 2007 cost bases. U.S. costs represent wellhead breakeven costs. Cost curves calculated assuming 10% real discount rate
Source: F. O'Sullivan, MIT Gas Supply Team analysis, ICF Hydrocarbon Supply Model, Data strictly for illustrative purposes only

10

Naturally, the variability in well productivity has major implications for the economics of the shale resource – Extensive drilling has pushed supply up and prices down, but much of this gas has been produced below cost

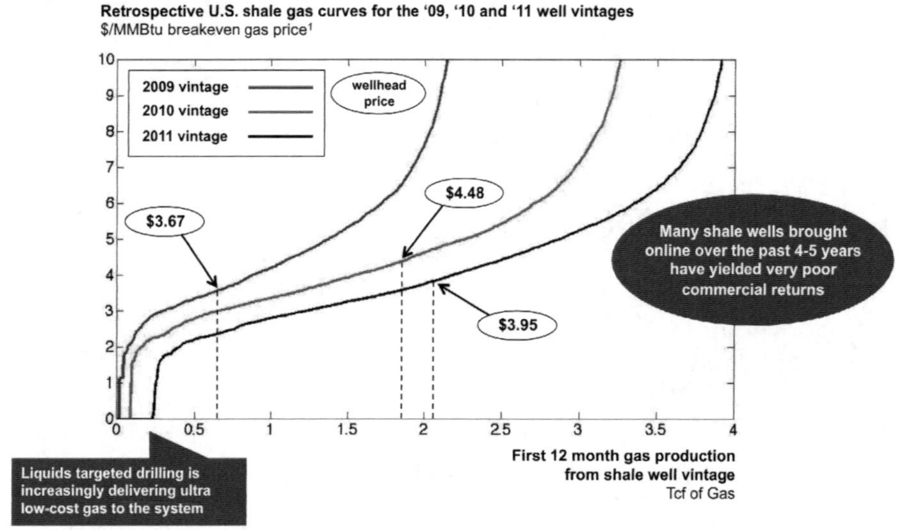

1. Supply curves include: Bakken, Barnett, Eagle Ford, Fayetteville, Haynesville, Marcellus and Woodford plays, and represent only gas produced by horizontal wells
Source: F. O'Sullivan

11

Domestic and international market evolution – The changing role for U.S. gas

12

Over the past several years falling gas prices have led to gas increasingly displacing coal-fired generation – More gas use for power generation going forward is certain

Evolution of natural gas prices for power generation from 2004 in the United States
$/MMBtu of gas

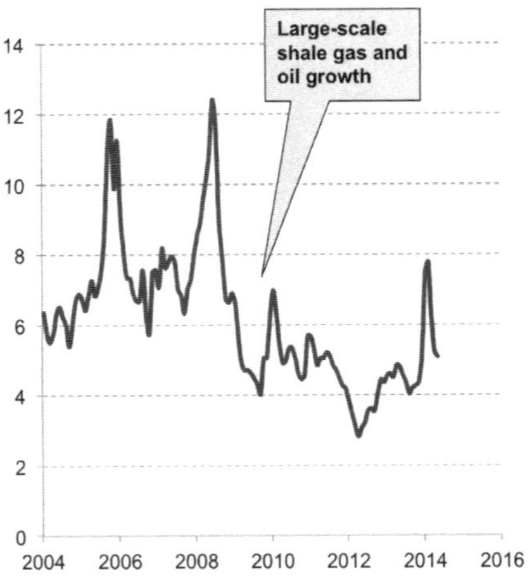

Comparison of coal and gas-fired power generation levels in the U.S. since January 2008
TWhrs

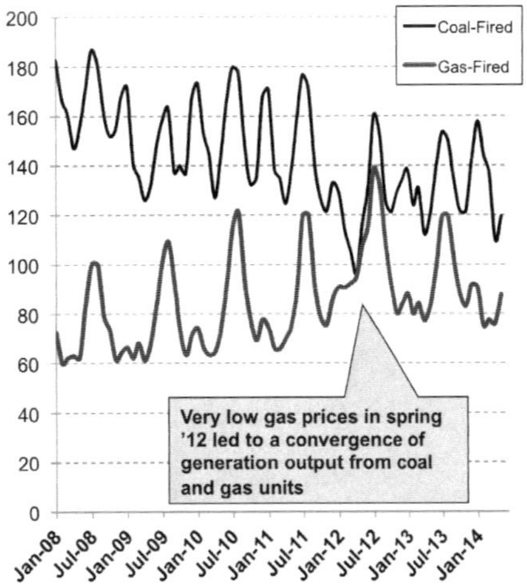

Source: F. O'Sullivan, United States Energy Information Administration

13

CO_2 emissions from U.S. power generation have fallen by 15% since 2005 due largely to coal-to-gas switching – The emission reductions have had negative costs, but how sustainability these reductions will be is unclear

US CO_2 emissions from electric power generation in 2005 and 2012
Tg of CO_2

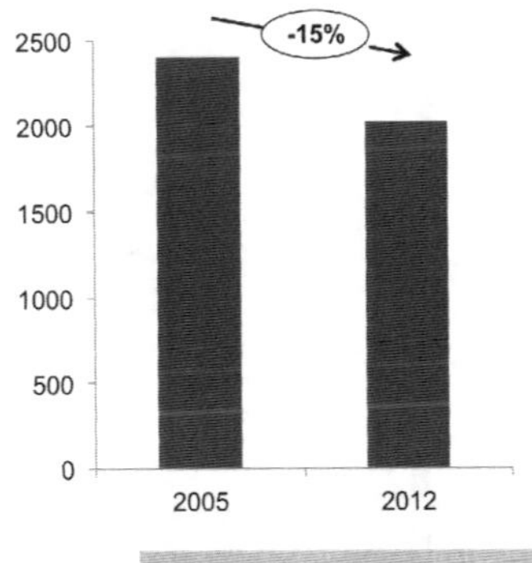

The McKinsey abatement curve - 2007

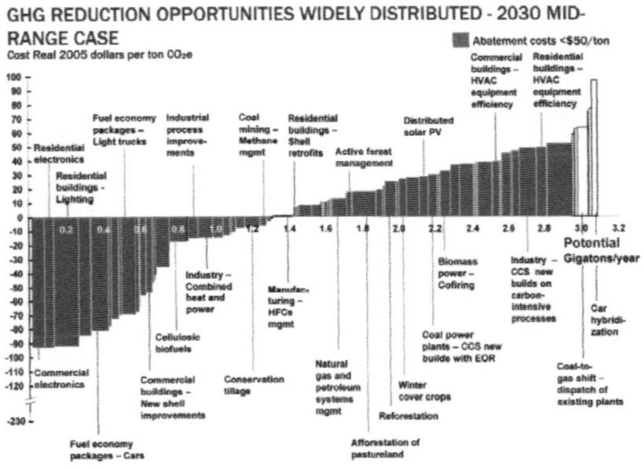

- In 2007, McKinsey estimated that coal-to-gas switching would yield 80MT of CO_2 abatement at at cost of >$50/ton
- Compared to 2005 levels, lower cost gas from shale has resulted in >350MT of annual CO_2 abatement, at zero to negative cost

Source: McKinsey & Company, United States Environmental Protection Agency

14

Going forward, the abundance of moderate-cost gas in the U.S. points to gas-fired generation dominating new build – This conclusion cannot be drawn for other regions where gas markets are oil-linked

Gas price needed for new build coal selection ahead of NGCC assuming various carbon emissions limits and no EOR[1]
$/MMBtu

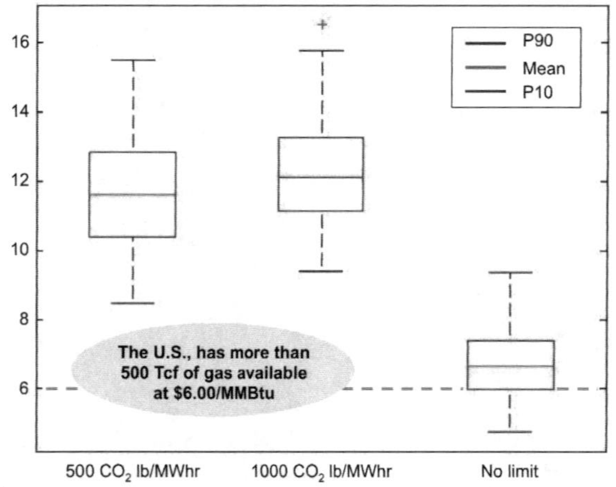

- **The structure of global gas markets means that dynamics in the U.S. are not necessarily relevant elsewhere**
- **U.S. gas-on-gas market will likely continue to ensure gas is the most attractive new build option**
- **Liquids-linked pricing outside North America means coal-fired plants are less disadvantaged**
- **The breaking of oil-linked gas pricing in East Asia over the coming years is a major uncertainty**

1. N=10,000. Heat rate, Capital cost and O&M costs scale linearly. CO2 emissions are function of emission standard, heat rate and unabated emissions. Costs based on NETL (2011) for PC, EIA (2011) and IEA (2011) for NGCC. Costs for NGCC are interpolated from reported costs on 0% and 90% capture. Natural gas price constant over lifetime of plant. Capacity factor: 75% with 5% standard deviation. Transport and storage cost: $15/ton CO2 captured. Capital charge: 15%. Coal price: $2.1/MMBtu. Capital cost uncertainty: if capture, standard deviation $100/kW, if no capture, standard deviation $25/kW.

Source: Analysis by J. Eide & H. Herzog, MIT, F. O'Sullivan

15

The global gas price differentials that exist today makes U.S. shale exports (to Asia) look very attractive – The U.S.-Japan differential will likely narrow over the coming years but shale gas supply will likely remain in-the-money

Variation in natural gas spot pricing in major global markets since 2000
$/MMBtu

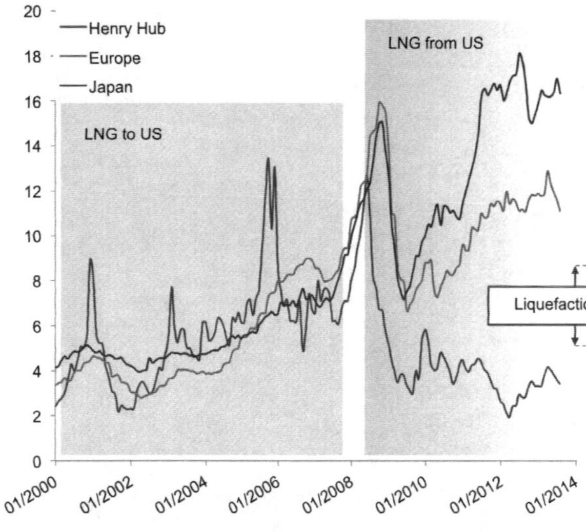

Conceptual U.S. shale-based LNG value chain

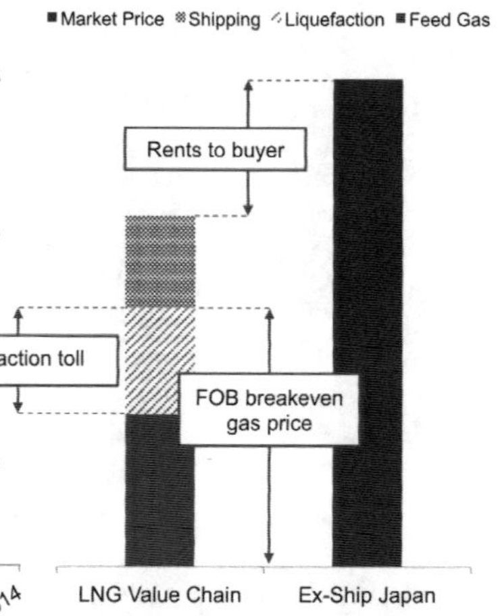

Source: F. O'Sullivan, EIA, Bloomberg

16

Considering the medium term supply stack, U.S. export projects will be quite competitive in supplying Asia – The real risk lies in the feed gas price, though most U.S. projects are passing this through and running as tolling operations

Estimation of new LNG supply curve to 2025 based upon project with high probability of completion
Ex-ship Tokyo breakeven price - $/MMBtu

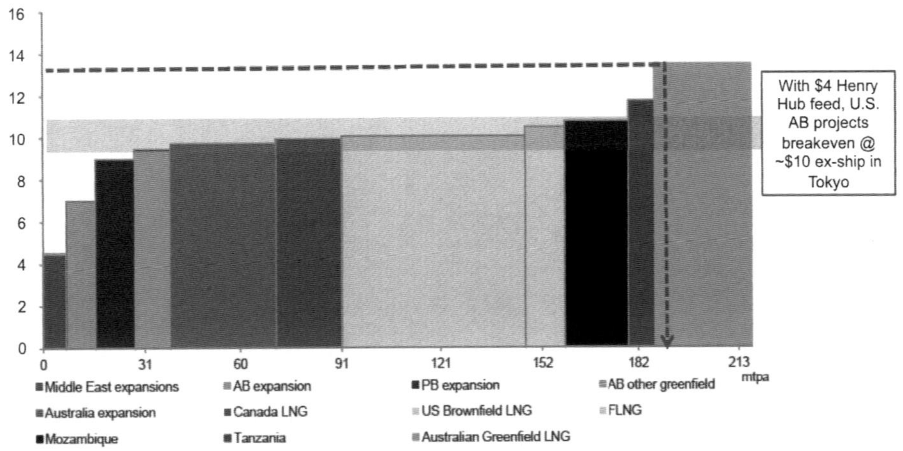

Source: F. O'Sullivan, IGU, Deutsche Bank

17

The capacity seeking LNG export approval is enormous but the realized levels will be more modest – At full capacity, current licenses for NFT LNG export would more than double U.S. gas exports

Snapshot of U.S. LNG import and export capacity as of September 2013
Bcf/day

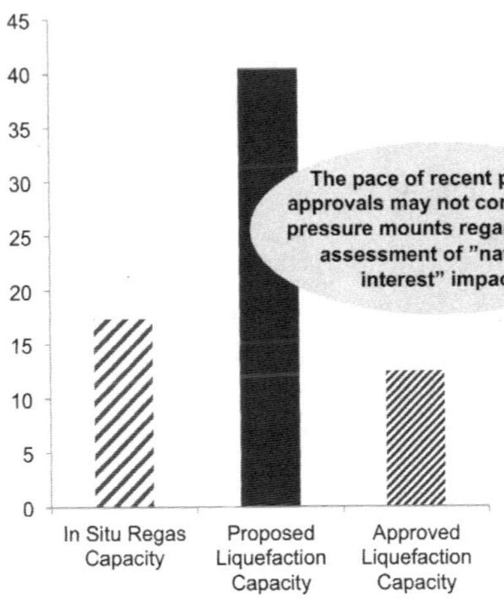

Variation in U.S. natural gas imports and exports from 2000 and 2013
Bcf/day

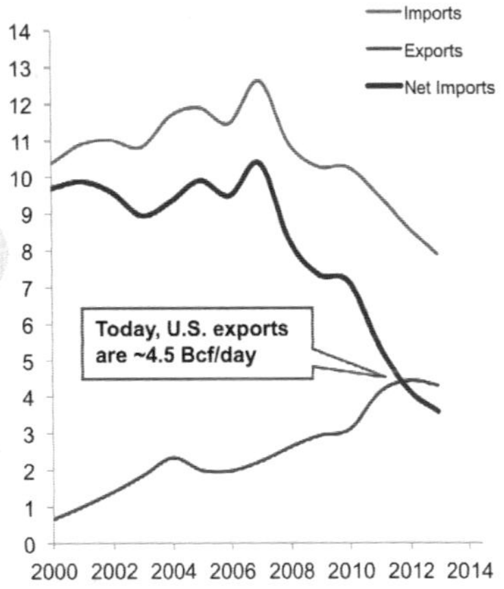

Source: United States Energy Information Administration, United States Department of Energy, CRS

18

Natural Gas Based Stationary Power Generation

Panel Discussion Agenda

- Introduction (10-15 minutes)
- Introduction and overview by panelists (each 5-7 minutes)
- Panel discussion starts (60 minutes)
- Open questions from audience (7-10 minutes)

Estimated U.S. Energy Use in 2012: 95.1 Quads
Contributions of Major Energy Sources

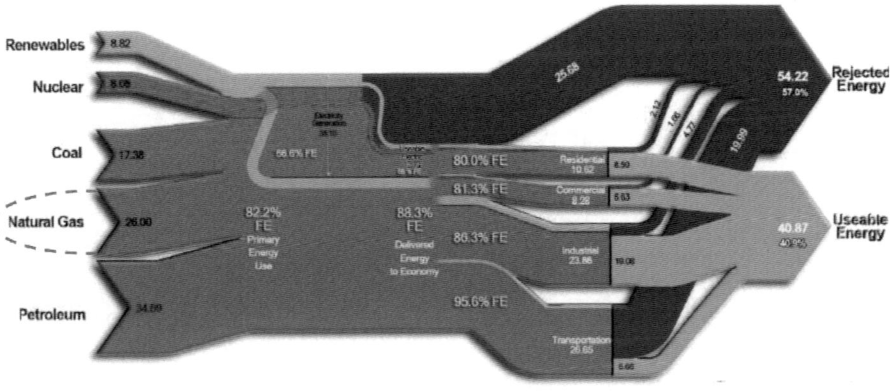

Source: DOE EIA-0035 (2013/05)

NG: The Rise of Production and Fall of Price

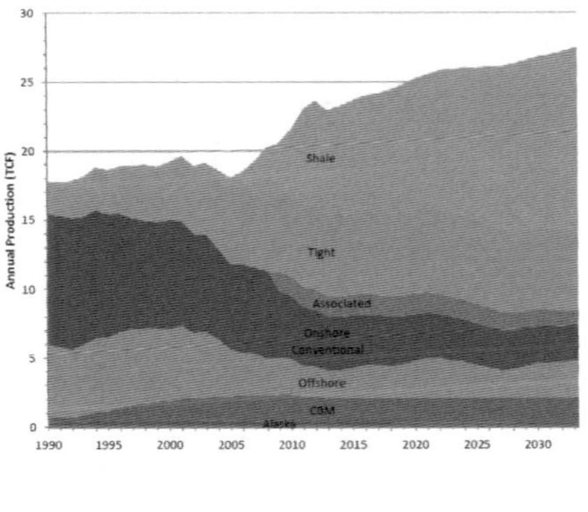

Source: DOE EIA-0035 (2013/05)

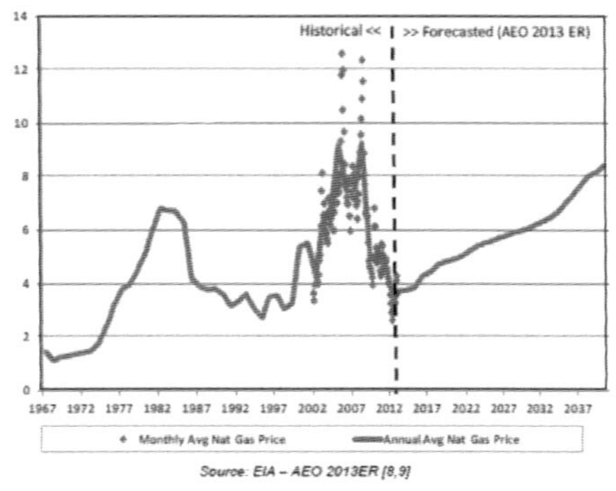

Source: DOE/NETL-341/061013

Power Plants Performance

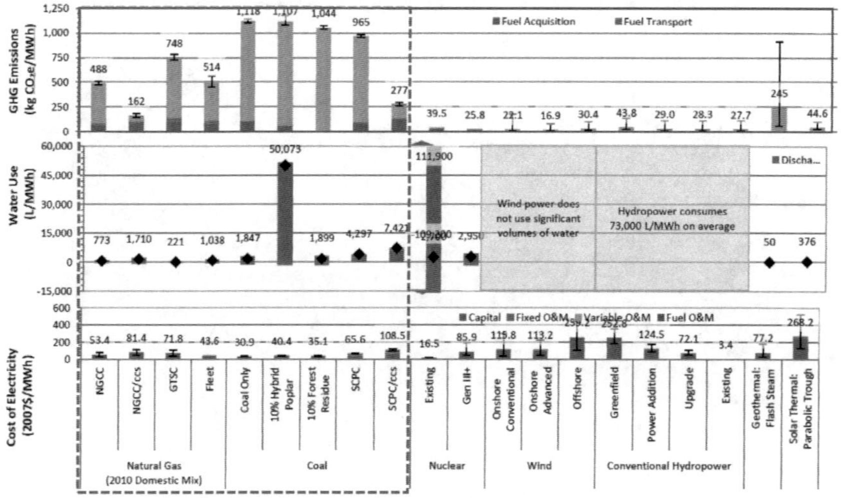

Source: DOE/NETL Fact Sheet, NETL program/product Identification: Life Cycle Analysis

Major NG Applications

- Power generation
  - Centralized: NGCC
  - Distributed: GTSC, NGFC, reciprocal engines, microturbines
- Chemical synthesis
  - H_2 production
  - Fertilizers
  - Refinery
  - Transportation fuels

Centralized NGCC

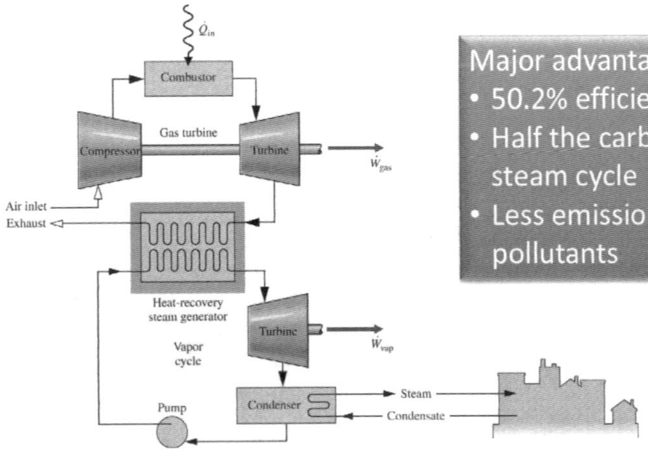

Major advantages:

- 50.2% efficiency
- Half the carbon emission of steam cycle
- Less emission of other air pollutants

Source: Moran et al, FUNDAMENTALS OF ENGINEERING THERMODYNAMICS, 7th ed., 2011

NGCC with CCS

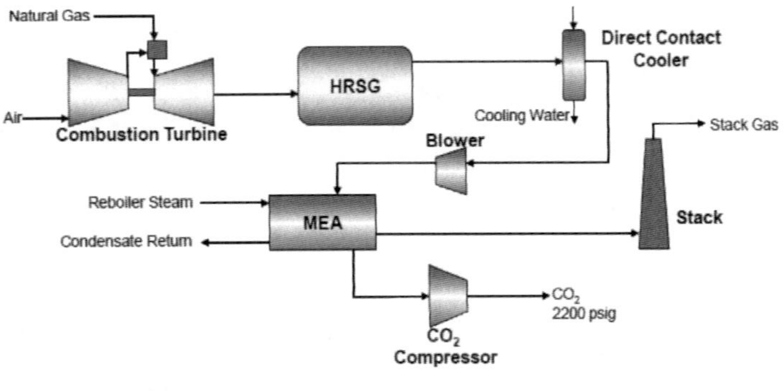

NOx Control: LNB + SCR to maintain 2.5 ppmvd @ 15% O_2
Steam Conditions: 2400 psig/1050°F/950°F

Source: Life Cycle Results from the NGCC LCI&C Study, NETL, September, 2012

NGSOFC for DG

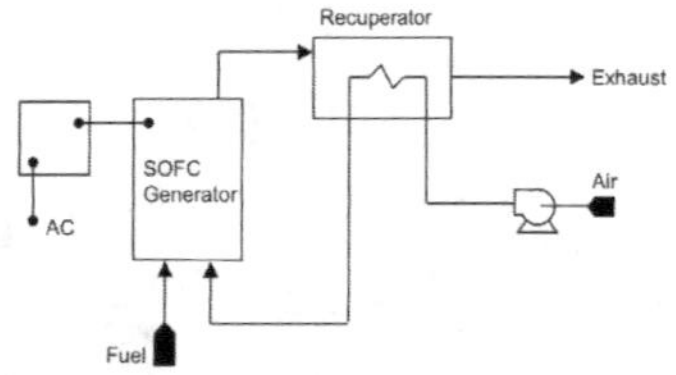

Atmospheric Pressure

- Efficiency: 45-50%
- Heat/Power Ratio= 0.8
- Exhaust temp: 600F
- Ideal for CHP

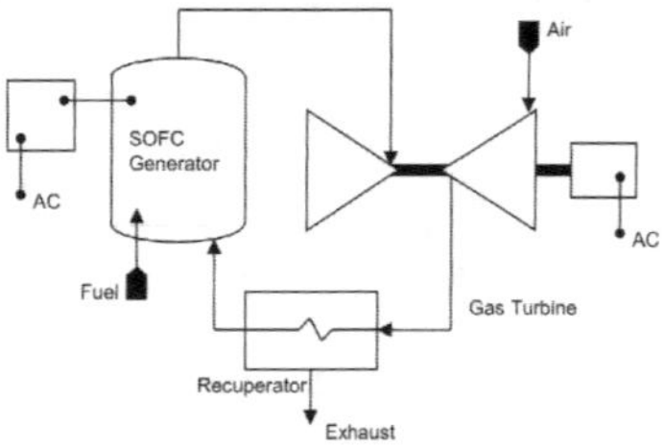

Pressurized Hybrid

- Efficiency: 55-60%
- Heat/Power Ratio= 0.5
- Exhaust temp: 500F
- Suitable for CHP

Westinghouse EDB/ELSAM 100 kW SOFC-CHP

100 kWe SOFC Power System Operation Summary

- 20,350+ hours
  - 4035 hours Pittsburgh/Netherlands in 1998
  - 12,577 hours in The Netherlands in 1999 & 2000 (95% availability)
  - 3,500+ hours at Essen, Germany
- Major performance:
  - 46% [Net ac/LHV] at 109 kWe AC to the Grid
  - 65 kWth to district heating system
  - NO_x <0.2 ppmv, undetectable SO_x, CO, VOCs
  - No detectable cell degradation

Siemens/Westinghouse 220 kWe Pressurized SOFC/Gas Turbine Hybrid System

Performance Summary

- 774+ hours
  - 110 hours FAT
  - 150 hours SAT + operation
  - 514 hours operation
- 188 kW total
  - SOFC - 166 kW
  - MTG - 22 kW
- 52% (net ac/LHV)

CCS Technology

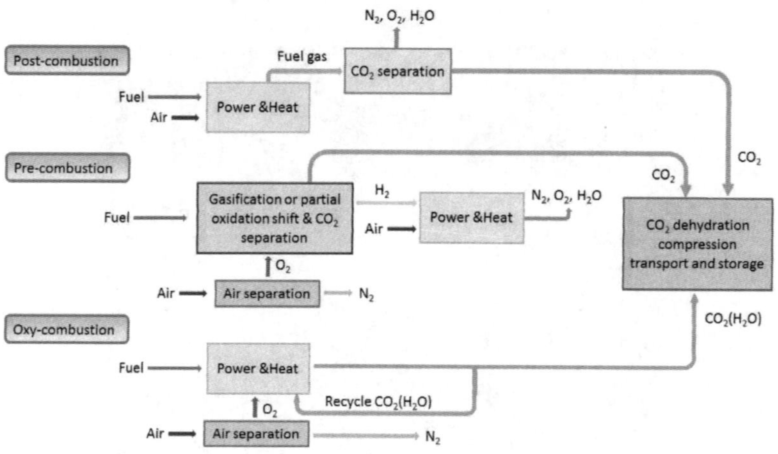

Carbon Capture Technologies

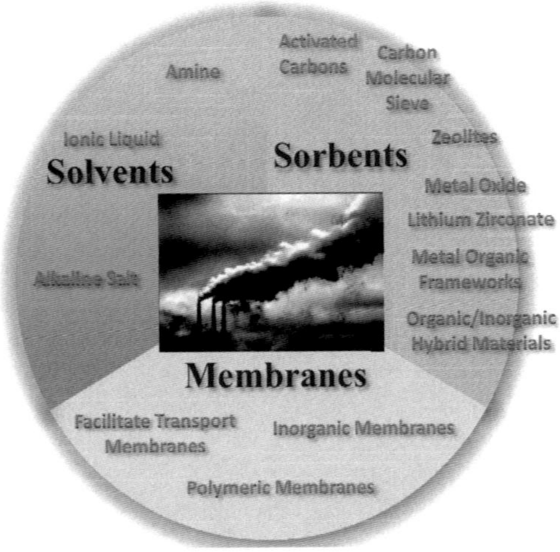

H_2 Production

SMR: 96% H_2 production in the US

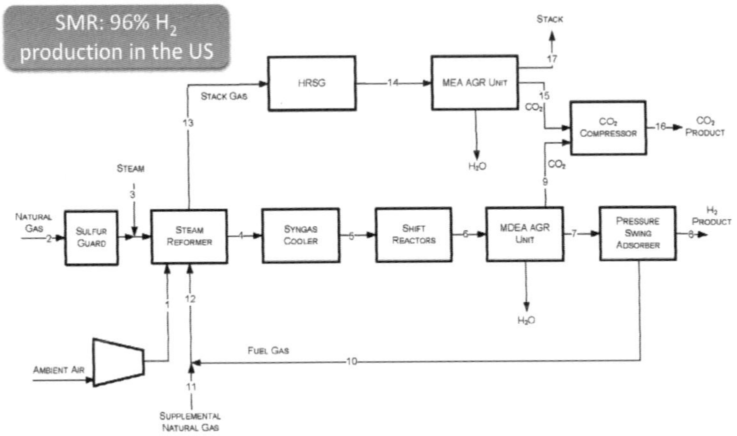

Source: DOE/NETL-2010/1434

H_2 Utilization

- Fuel cells
- Fertilizers
- Refineries
- Transportation fuels

Barriers to Implementation

- NGCC
  - High-temperature corrosions
- NGSOFC
  - Cost, durability and performance
- CCS
  - Cost, energy and efficiency penalty
- HCU
  - Cost and performance

Questions to be Discussed

- What are *potential areas for expansion* for NG in stationary energy generation?
- What are the major *technical, regulatory, or other barriers* or roadblocks that prevent growth of NG in stationary energy generation?
- What are the primary *technical gaps* inhibiting wider NG utilization?

Questions to be Discussed

- What science or engineering research and development programs should be expanded to address these gaps?
- What is the magnitude or the *level of impact of the technology gaps*?
- What *role should the Government (i.e., federal and state) play* in addressing potential technology or policy gaps?

Backup Slides

Westinghouse Field Test Units

Time Year	Customer	Stack Rating (kWe)	Cell Length (cm)	Cell Number	Oper. (Hrs.)
1986	TVA	0.4	30	24	1760
1987	Osaka Gas	3	36	144	3012
1987	Osaka Gas	3	36	144	3683
1987	Tokyo Gas	3	36	144	4882
1992	JGU-1	20	50	576	817
1992	UTILITIES-A	20	50	576	2601
1992	UTILITIES-B1	20	50	576	1579
1993	UTILITIES-B2	20	50	576	7064
1994	SCE-1	20	50	576	6015
1995	SCE-2	27	50	576	5582
1995	JGU-2	25	50	576	13,194
1998	SCE-2/NFCRC	27	50	576	5380+
1997	EDB/ELSAM - 1	100	150	1152	4035
1999	EDB/ELSAM - 2	100	150	1152	12,577
2001	RWE 100kW	100	150	1152	3500+
2000	SCE	200	150	1152	774+

3 MW SOFC/GT Hybrid - 70% Electrical Efficiency

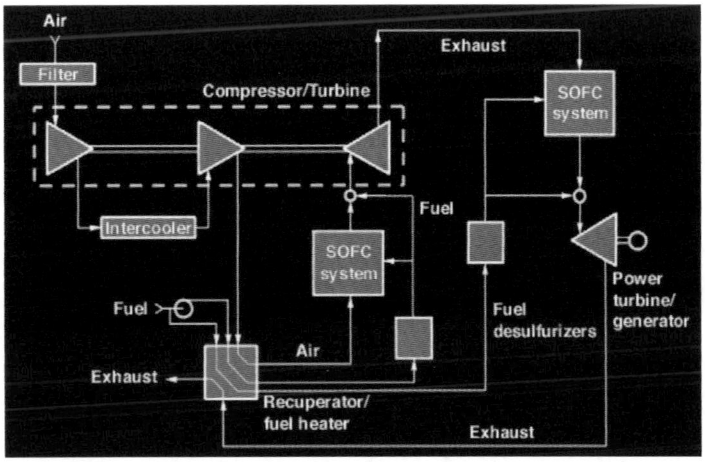

3 MW Power System Layout

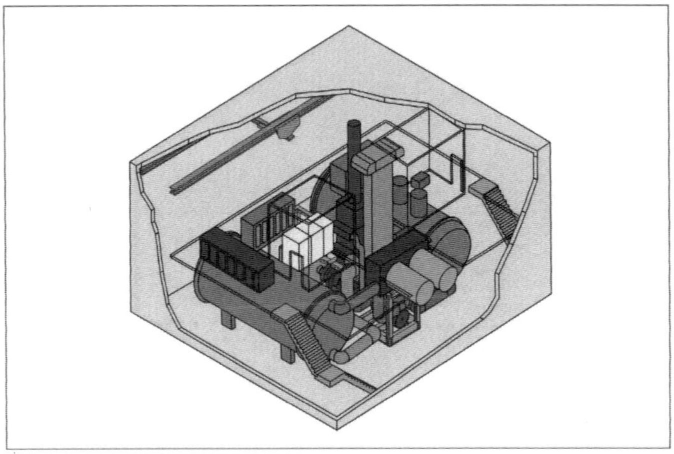

300 kWe-Class Hybrid PSOFC

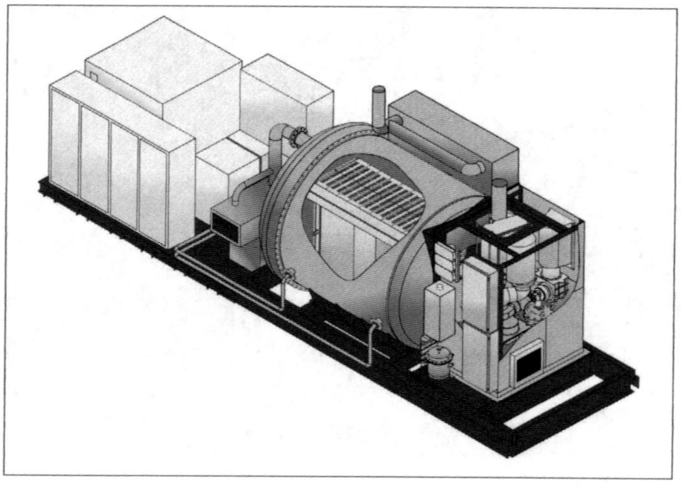

500 kW SOFC/GT Hybrid System

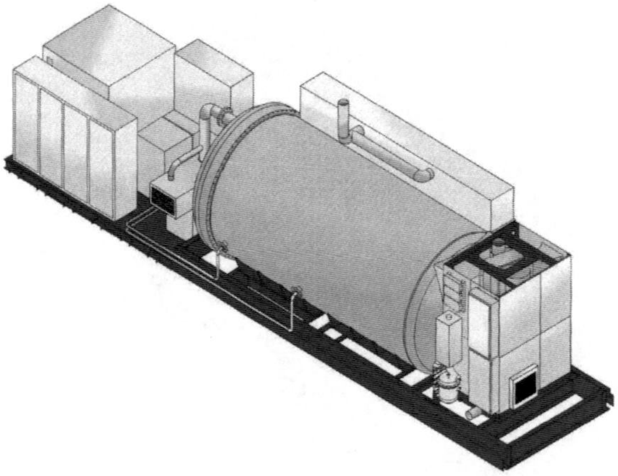

NGFC Configuration – Ambient Pressure SOFC

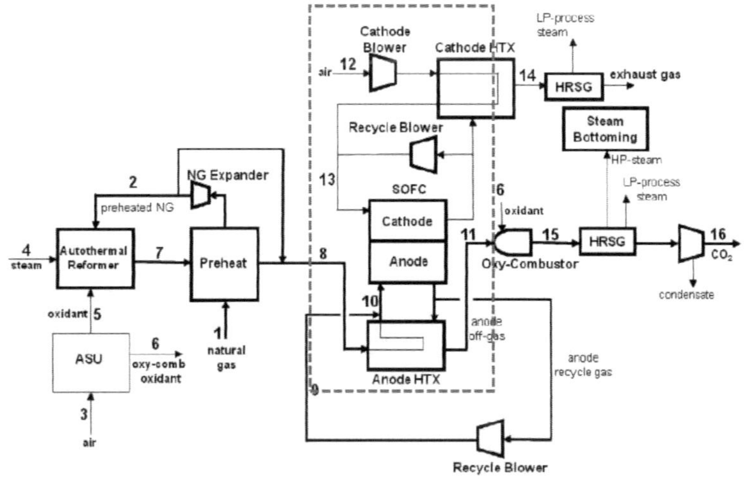

Source: DOE/NETL – 2011/1486

NGFC Configuration – Pressurized SOFC

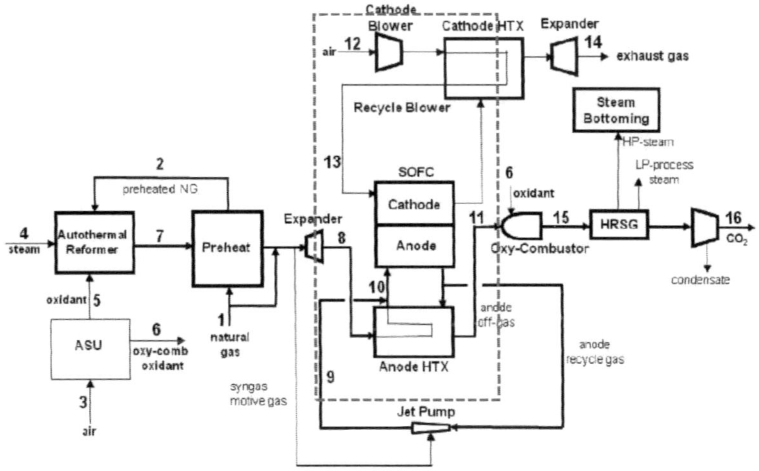

Source: DOE/NETL – 2011/1486

NGFC Performance

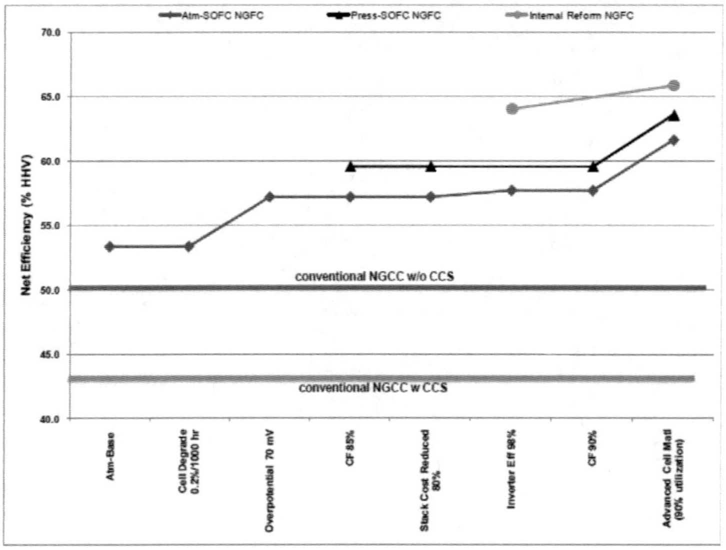

Source: DOE/NETL – 2011/1486

NGFC Performance

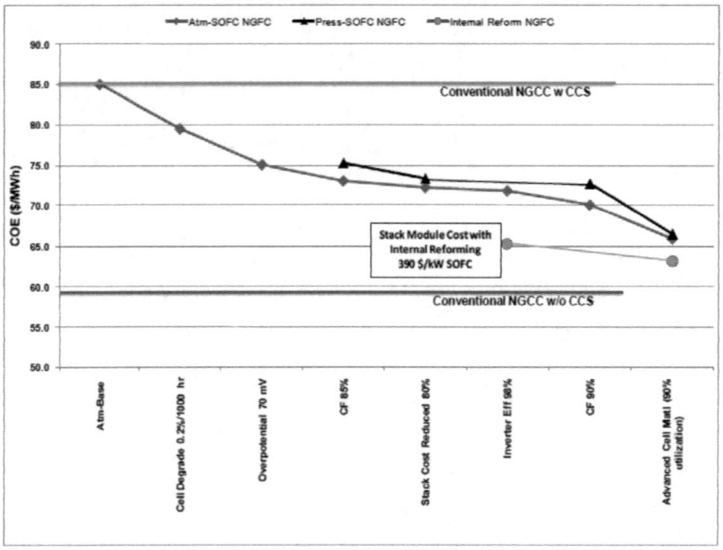

Source: DOE/NETL – 2011/1486

SOFC for DG

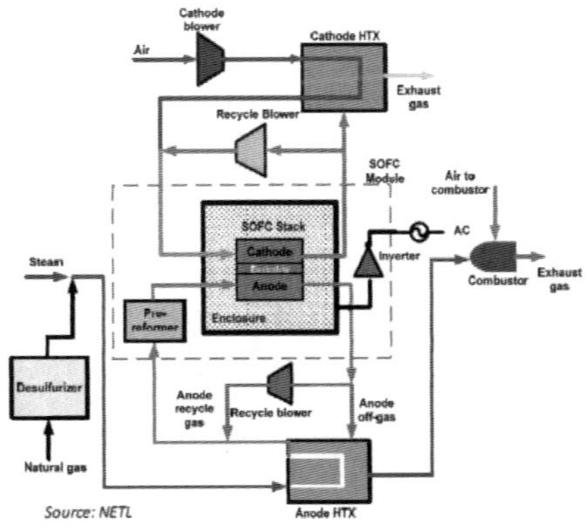

Source: NETL

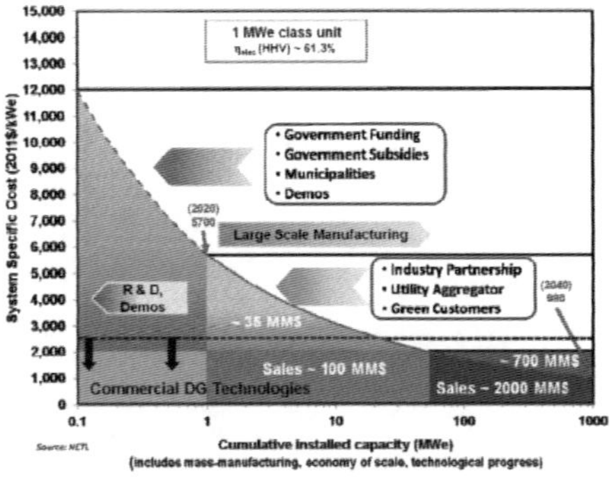

Source: DOE/NETL- 342/093013

Air Products: Natural Gas Conversion to H_2

- Air Products: World's leading supplier of Merchant H_2, Leading supplier of industrial gases such as O_2
- Today: Majority of hydrogen is made from natural gas
  - Steam-methane reforming: $H_2O + CH_4 + \text{heat} \rightarrow CO + 3H_2$
    - 60% of CO_2 could be readily captured. Remainder lost in low pressure flue gas
- Today: Oxygen made by cryogenic distillation
- Future: Ceramic Membranes: ITM Syngas
  - Up to 90% of CO_2 could be captured
- Future: Ceramic Membranes:
  - ITM Oxygen
  - Polygen to make oxygen, power and steam

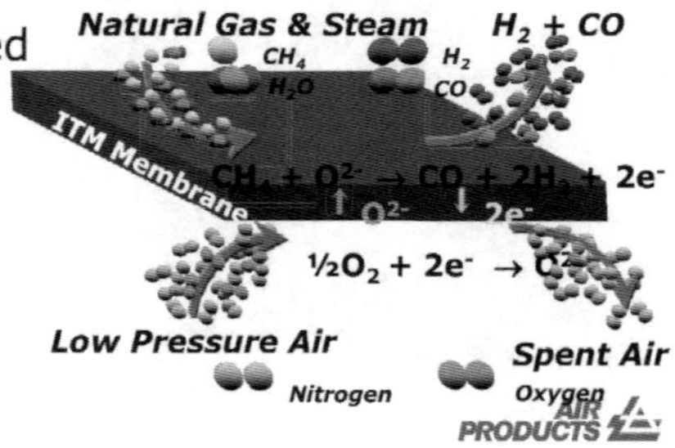

AIR PRODUCTS

Technical gaps that could be addressed by the DOE and the National Labs:

- Improved ceramic membranes:
  - improved materials
  - membrane designs
  - fabrication methods
- Basic ceramic science and engineering of the application of large scale, ceramics components in demanding service
- Development of enabling technologies for the balance of plant
- Larger scale demonstration projects

AIR PRODUCTS

Presentation to AIChE Alternative Natural Gas Applications Workshop
Panel on stationary energy generation
Alexandria, Virginia
October 8, 2014

GAS TURBINE ASSOCIATION

Competitiveness of US Gas Turbine Manufacturers

William H. Day
Managing Director, Gas turbine Association

October 8, 2014 1

- Alstom
- GE Energy
- Florida Turbine Technologies
- Meggitt Sensing Systems
- Pratt & Whitney
- OPRA Turbines
- PW Power Systems
- Rolls-Royce Energy Systems
- Siemens Energy
- Solar Turbines

October 8, 2014 Page 2

Acknowledgement

The information on the following charts is from a study commissioned by the Gas Turbine Association and performed by ICF International, 3/29/13.

To see the full study, go to the GTA website www.gasturbine.org and click on the "Hot Topics" tab. Then click on "GTA White Paper on R&D Funding".

October 8, 2014 Page 3

Background

- US-based manufacturers currently have a large share of the simple cycle and combined cycle market.
- With the expansion of European mfrs to the US, the only other major producer of heavy frame units in the US is Japan.
- Due to the Fukushima disaster, the Japanese government is funding large R&D to improve CC efficiency as an alternative to decreased nuclear power.
- Without an expansion of US government R&D funding, US-based mfrs will lose their competitive advantage in the US electricity market.

October 8, 2014 Page 4

Results Summary

- Foreign mfrs will gain a large share of the US market for new electric generating capacity over the next 25 years unless US mfrs develop advanced NGCCs.
- The study forecasts an increase in US jobs with new capacity expansion.
- If US-based NGCC efficiency targets are not met and foreign mfrs meet theirs, there will be fewer US jobs as GT mfg is increasingly done offshore.
- Developing high-efficiency NGCCs will reduce the growth of CO_2. AS US NGCCs will be more efficient than foreign ones, other countries will produce less CO_2 as well.

October 8, 2014 Page 5

Projected Combined Cycle Build Out

Exhibit ES-1

Projected Combined Cycle Build Out

CCs	Actual as of 2013		2037 - Foreign government investment Case		2037 – Enhanced US investment Case	
	MW	%	MW	%	MW	% of Total
US	191,100	84%	191,900	42%	327,745	72%
Non-US	35,196	16%	262,432	58%	128,371	28%
Total	**226,296**	**100%**	**454,333**	**100%**	**456,115**	**100%**

Source: Ventyx and IPM output

- In the Foreign government investment case, without increased US government R&D funding US mfrs will become much less competitive.
- Market share would drop to 42% from 84% today.

October 8, 2014 Page 6

Projected Jobs Impact

Exhibit ES-2
Projected Jobs Impact on the US Economy

Select Years	Foreign government investment Case	Enhanced US investment Case	Delta
2020	26,000	23,000	-3,000
2030	16,5000	203,000	38,000
2035	172,000	208,000	36,000
Annual Average	90,000	102,000	12,000

Source: IMPLAN output

The Enhanced US Investment case supports 36,000 (~ 20%) more jobs vs. the Foreign Government case by 2030.

October 8, 2014 Page 7

Timeline of Advanced Combined Cycles

Exhibit 1-1
Rollout Timeline of Advanced Combined Cycles

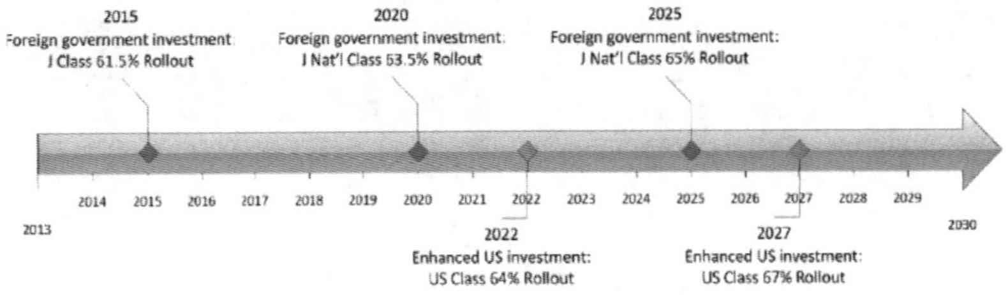

- Funding needed for advanced NGCCs: $50 million per year for 7 years
- Representative Tonko (D-NY) introduced bill 112H2781 on 5/31/13 to provide for the R&D funding

October 8, 2014 Page 8

GLOBAL
CCS
INSTITUTE

GLOBAL CCS INSTITUTE

Ron Munson, Principal Manager Carbon Capture
Alternative Natural Gas Applications Workshop
October 8, 2014

GLOBAL CCS INSTITUTE

Fossil fuels will be important for a long time to come

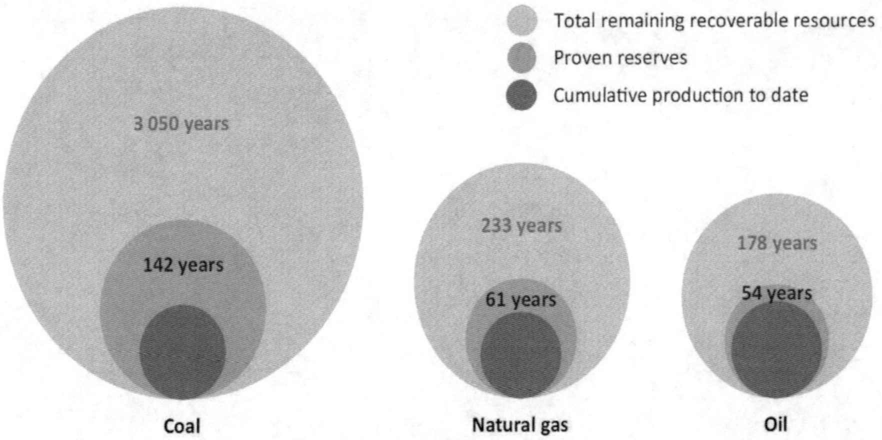

Source: IEA (2013)

134 GW coal capacity added in 2013 – at least double that of any other fuel. – IEA 2014

When expectations collide

- 'Business as usual' is incompatible with climate change objectives.
- 'Business as usual' leads to a world of rising average global temperature.

> *A 4°C world is so different from the current one that it comes with high uncertainty and new risks that threaten our ability to anticipate and plan for future adaptation needs.*
>
> World Bank Group President Jim Yong Kim, November 2012

- Renewable technologies are not the sole answer to mitigating the rise in global temperature:
  - decarbonizing power without CCS would cost 40% more than with CCS
  - in many industrial processes, accounting for over 20% of CO_2 emissions, there is no alternative to CCS.

CCS is a critical component of a sustainable energy system

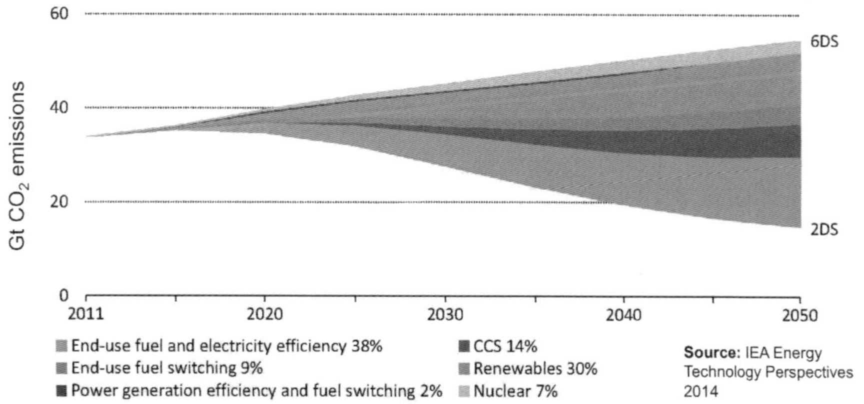

In a 2° scenario, CO_2 captured in 2030 is in excess of 1500 Mt and 6300 Mt in 2050.

No CCS – No 2°C

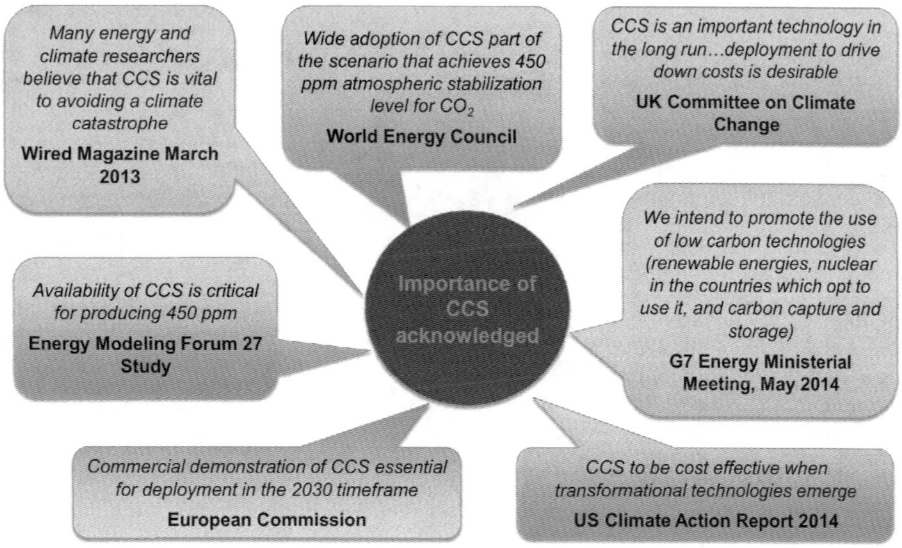

Government and Industry Advance Carbon Pricing

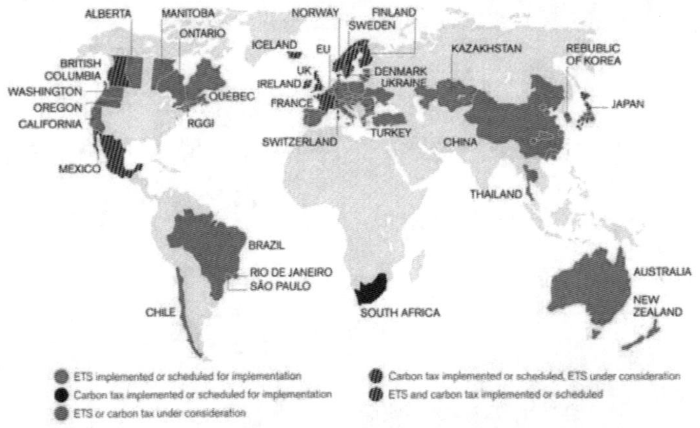

Source: World Bank (2014)

39 National and 23 Sub-national Jurisdictions – World Bank 2014;
150 Major Companies assign "shadow price" – CDP - 2014

Business case for CCS

TECHNOLOGY

- Confidence that the technology will work
- Operational projects in power critical to positive perception of CCS and to gain experience.

POLICY AND MARKETS

- Importance of CCS consistently acknowledged in global climate discussions, but
- CCS is not treated equivalently to other low-carbon technologies.

UNDERSTANDING AND ACCEPTANCE

- Awareness and understanding of CCS is low, perception of it as 'experimental'
- Operational projects are key to turn this around.

Collaboration and aligning three pillars of the business case for CCS

GLOBALCCSINSTITUTE.COM

FuelCell Energy
Ultra-Clean, Efficient, Reliable Power

Fuel Cells and Hydrogen for Greater Sustainability Using Natural Gas

Pinakin Patel

At AIChE Workshop. Washington DC October 8-9 2014

Ultra-Clean | Efficient | Reliable Power

FuelCell Energy
Ultra-Clean, Efficient, Reliable Power

FCE Products and Services

Design
Megawatt–class distributed power generation solutions

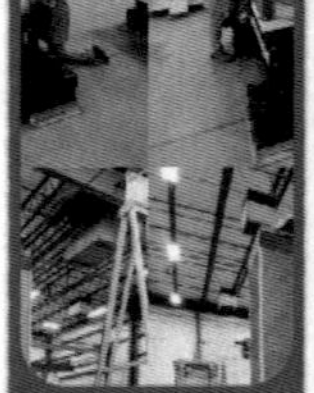

Manufacture
Global manufacturing footprint

- *North America*
- *Europe*
- *Asia via partner*

Sales
Direct & via Partners

Installations & orders in 9 countries

EPC*
Project Development and Project Finance, Engineering & Construction

Over 300 megawatts installed and in backlog

* Engineering, Procurement & Construction

Services
Operate & maintain power plants

- *Over 100 DFC® plants operating at more than 50 sites globally*
- *Three billion kWh ultra-clean power produced*

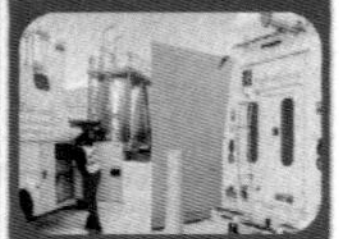

Providing turn-key distributed power generation solutions that meet both economic and sustainability goals

2

FuelCell Energy
Ultra-Clean, Efficient, Reliable Power

Scalable Solutions

Individual fuel cell & 350 kW fuel cell stack

Four-Stack Module 1.4 megawatts

Completed module 1.4 megawatts

1.4 MW DFC1500®
- *Utilizes one module*
- *Adequate to power 1,400 homes*

2.8 MW DFC3000®
- *Utilizes two modules*
- *Adequate to power 2,800 homes*

59MW fuel cell park

3

FuelCell Energy
Ultra-Clean, Efficient, Reliable Power

World's Largest Fuel Cell Park

"The scale of this installation is contributing to the power and heating needs of an urban population and generating the electricity in a ***highly efficient and ultra-low emission*** *profile that supports our National renewable portfolio standard,"*
Tae-Ho Lee
Chief Executive Officer
Gyeonggi Green Energy

- **Scalable consisting of 21 DFC3000® power plants**
  - Only ~ 5.2 acres for 59 MW
- **Supplying electric grid and district heating system**
- **Constructed in only 14 months**
- **Adequate to power ~ 140,000 S. Korean homes**

4

FuelCell Energy
Ultra-Clean, Efficient, Reliable Power

Distributed Hydrogen Production

Industrial Hydrogen Use
Existing market

Vehicle Fueling
Emerging market

Orange County Sanitation District
Renewable Hydrogen for Vehicle fueling

Ongoing Projects:

- Orange County Sanitation District, CA – Hydrogen from wastewater treatment digester gas for vehicle fueling
- Village Farms, Vancouver Canada - Hydrogen from landfill gas for vehicle fueling
- Torrington DFC Manufacturing Plant – Hydrogen for industrial heat treating

Zero or low-carbon H_2 economically produced near end users

5

FuelCell Energy
Ultra-Clean, Efficient, Reliable Power

What Can We Do With By-Product Hydrogen?

	DFC300®	DFC1500®	DFC3000®
Co-product			
Power, kW	250	1,150	2,300
Hydrogen, kg/day	**125**	**700**	**1,400**
Heat, mmBtu/hr	0.5	2.0	4.0
Refueling Capacity			
Cars, 4.2 kg/day	**30**	**140**	**280**
Buses, 25 kg/day	**5**	**24**	**48**
Fork Lifts, 2.1 kg/day	**60**	**280**	**560**
Plug-in Battery Hybrid, 12 kWh/day	**500**	**2,400**	**4,800**

MO3256A

6

FuelCell Energy
Ultra-Clean, Efficient, Reliable Power

Enabling Renewable Energy Use for Grid Independent Operation

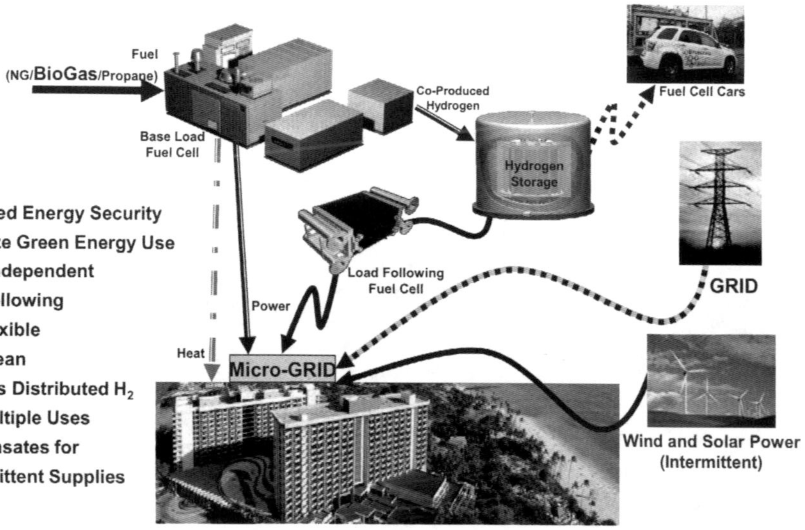

- **Enhanced Energy Security**
- **Maximize Green Energy Use**
- **Water Independent**
- **Load Following**
- **Fuel Flexible**
- **Ultra Clean**
- **Provides Distributed H_2 for Multiple Uses**
- **Compensates for Intermittent Supplies**

7

FuelCell Energy
Ultra-Clean, Efficient, Reliable Power

Chemical Integration / Hydrogen Recycle

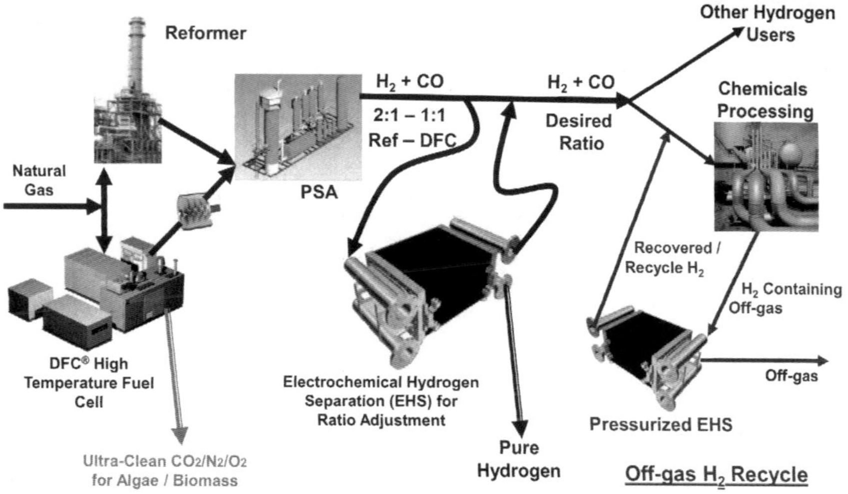

Chemical integration opportunities include H_2 to CO ratio adjustment, H_2 recycle from processing off-gas and CO_2 for other uses

8

FuelCell Energy
Ultra-Clean, Efficient, Reliable Power

Carbon Capture with DFC Powerplants

- Carbonate electrochemical process transfers CO_2 from Air Electrode (Cathode) to Fuel Electrode (Anode)
- CO_2 is easily separated from Fuel Electrode exhaust gas because it is no longer diluted with air
- Demonstrated in small cells, 9 kW test starting this year

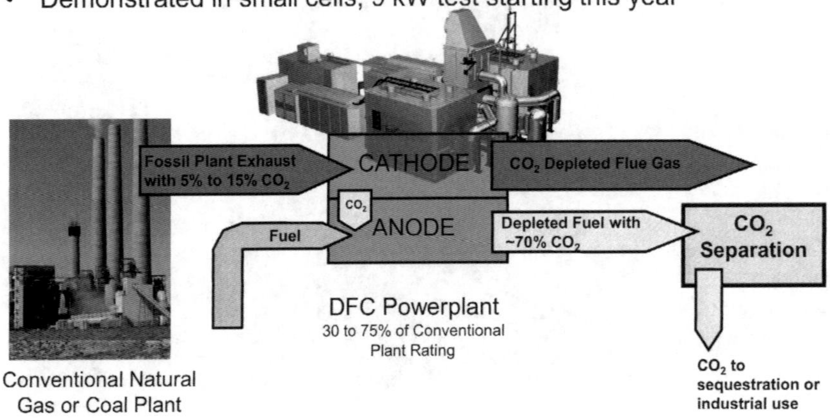

9

FuelCell Energy
Ultra-Clean, Efficient, Reliable Power

Application Examples

Back-up slides

10

FuelCell Energy
Ultra-Clean, Efficient, Reliable Power

Wind Power is Soft Energy

Hydrogen can also reduce size of battery storage needed by increasing average wind + battery + H_2 output

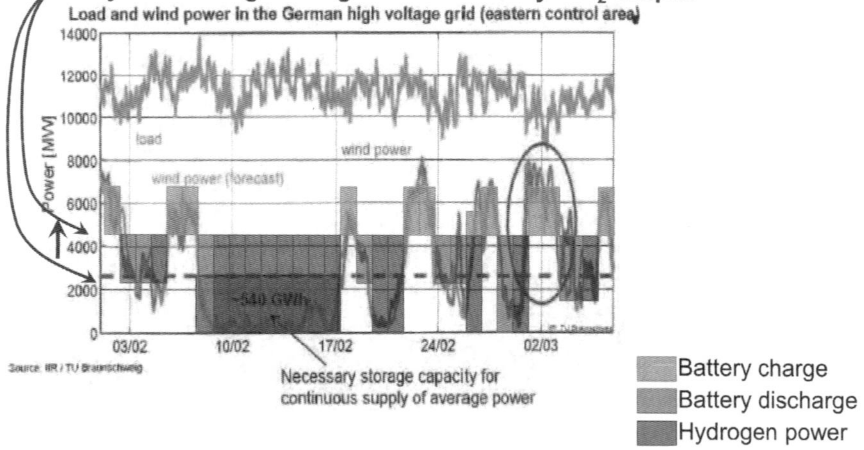

Hydrogen converts soft to hard, reliable power

11

FuelCell Energy
Ultra-Clean, Efficient, Reliable Power

Application Examples

Megawatt-class distributed power generation solutions

- **Highly Efficient**
  - High electrical efficiency
  - Combined heat & power supports economics and sustainability
- **Ultra-clean**
  - Virtual lack of pollutants eliminates future clean air compliance concerns
  - Low carbon footprint / net-zero carbon using renewable biogas
- **Enhances grid resiliency**
  - Continuous power close to the point of use

12

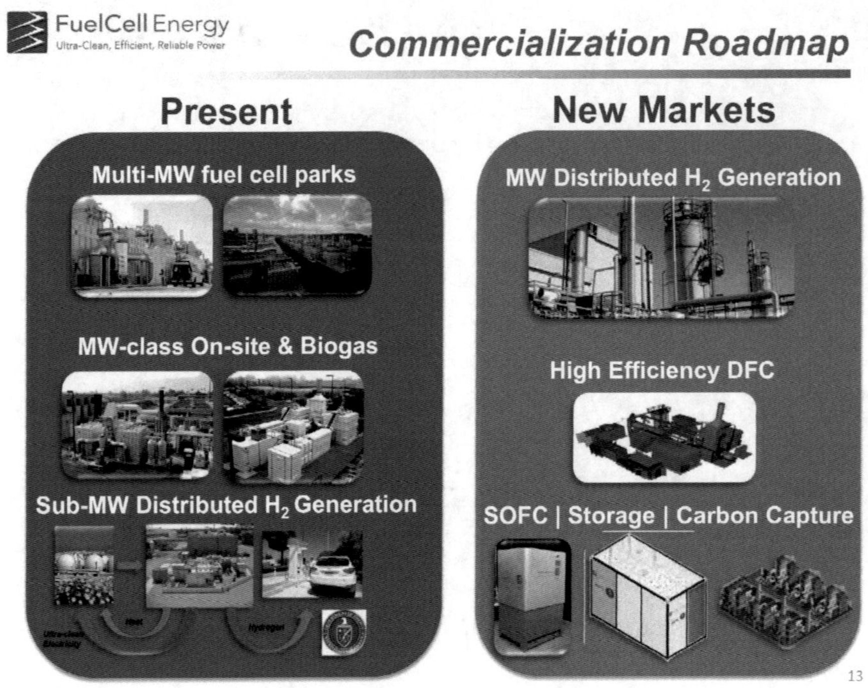
FuelCell Energy
Ultra-Clean, Efficient, Reliable Power
Commercialization Roadmap
Present
New Markets
Multi-MW fuel cell parks
MW-class On-site & Biogas
Sub-MW Distributed H_2 Generation
MW Distributed H_2 Generation
High Efficiency DFC
SOFC | Storage | Carbon Capture
13

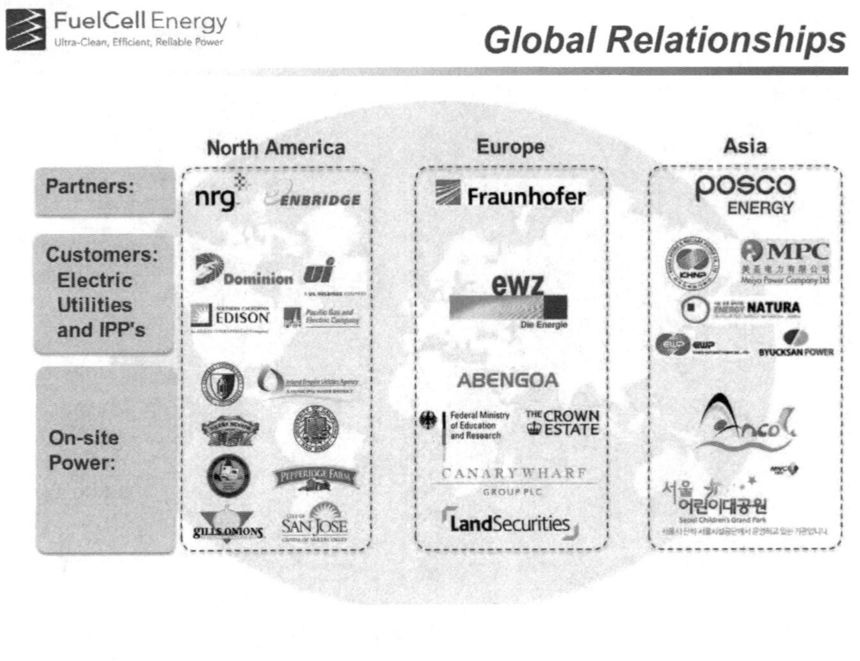
FuelCell Energy
Ultra-Clean, Efficient, Reliable Power
Global Relationships
North America
Europe
Asia
Partners:
Customers: Electric Utilities and IPP's
On-site Power:
nrg
ENBRIDGE
Dominion
ui
EDISON
PEPPERIDGE FARM
GILLS ONIONS
SAN JOSE
Fraunhofer
ewz
Die Energie
ABENGOA
Federal Ministry of Education and Research
THE CROWN ESTATE
CANARY WHARF
GROUP PLC
LandSecurities
POSCO
ENERGY
MPC
NATURA
BYUCKSAN POWER
Ancol
서울 어린이대공원
14

FuelCell Energy
Ultra-Clean, Efficient, Reliable Power

Global Business

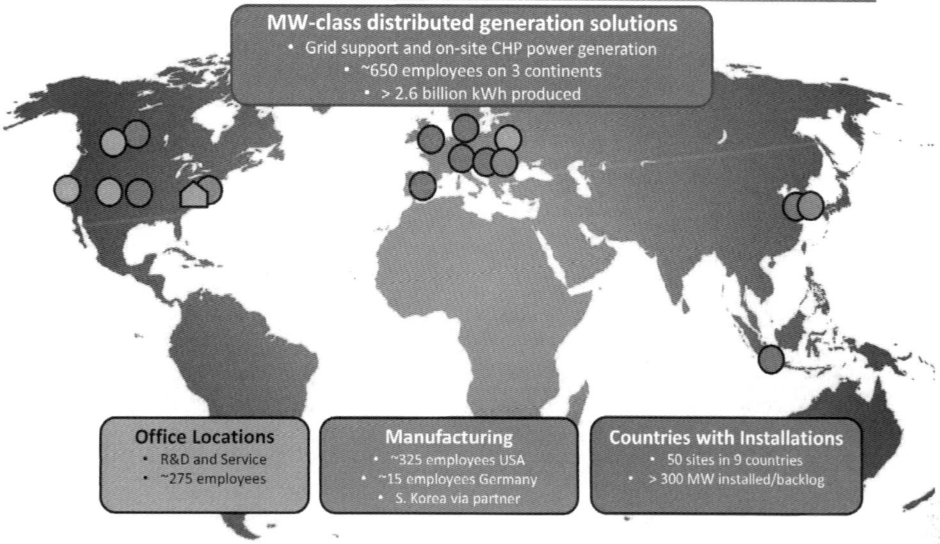

15

FuelCell Energy
Ultra-Clean, Efficient, Reliable Power

Competitive LCOE

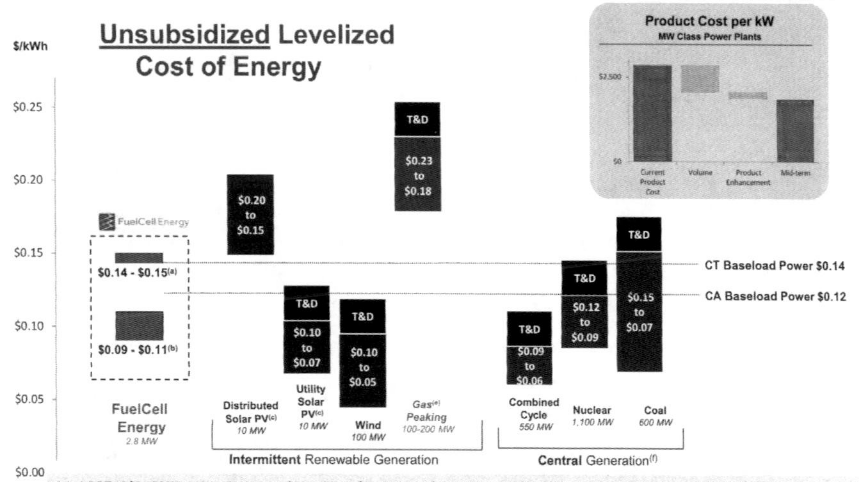

(a) LCOE of $0.15/kWh with natural gas at $8/mmBtu or $0.14.kWh at $6/mmBtu; each $2/mmBtu change equates to about $0.01/kWh.
(b) Mid-term LCOE target of $0.09-$0.11/kWh based on global production volume of approximately 210 MW annually.
(c) Distributed solar based on rooftop installation in SW USA with 20-23% capacity factor; Utility solar based on tracking technology and 27-28% capacity.
(d) Installation and maintenance cost of Transmission & Distribution (T&D) is estimated to add up to $0.024/kWh.
(e) Gas peaking addresses intermittency of solar and wind when power is required but sun not shining/wind not blowing.
(f) Does not include waste disposal costs, incremental emission clean-up costs or nuclear-related security costs.

Source: Company estimates, Lazard's Levelized Cost of Energy Analysis—Version 7.0 , U.S. Energy Information Administration (EIA) & Oak Ridge National Lab.

16

FuelCell Energy
Ultra-Clean, Efficient, Reliable Power

On-site Power & Heat

Project Overview

1.4 MW combined heat & power fuel cell plant located at Central Connecticut State University

- Providing ~1/3 of campus power needs

1.4 MW utility-owned power plant at CCSU – New Britain

Benefits

- High efficiency drives favorable economics
  - CHP for heating and absorption chilling
- Ultra-clean emission profile supports CCSU sustainability goals
- Micro-grid enhances campus energy security
- Private capital providing public benefits

"This power plant is a significant step in helping CCSU achieve our aggressive goals for reducing greenhouse gases and improving energy efficiency on campus," said ***Central Connecticut State University President Jack Miller.***

17

FuelCell Energy
Ultra-Clean, Efficient, Reliable Power

Electric Grid Support

Project Overview

14.9 MW of ultra-clean, efficient and continuous power generation on a remediated brownfield site in Bridgeport, CT

Output is adequate to power approximately 15,000 homes

Benefits

- Clean & quiet on ~ 1.5 acres
- Hardens electric grid by supplying 3 substations
- Tax revenue for City of Bridgeport
- Sustainable jobs for State of Connecticut
- 15 year service contract for FuelCell Energy
- Diversify power generation for utilities/consumers

Dominion

ui The United Illuminating Company

CL&P

"The Dominion Bridgeport Fuel Cell is another important step in our efforts to identify and develop opportunities to produce clean energy that is reliable and cost effective," said ***Thomas F. Farrell II, Dominion Chairman, President and Chief Executive Officer.*** *"We are now adding fuel cell technology to our energy portfolio that already includes wind, hydro, biomass and, soon, solar. This project supports Connecticut's clean energy goals while producing significant economic development benefits for the State and the City of Bridgeport."*

18

FuelCell Energy
Ultra-Clean, Efficient, Reliable Power

Renewable Power

Challenge for a Water treatment facility

- Compliance with clean air regulations
- Meet ambitious sustainability goals

Solution

World's largest fuel cell power plant converting renewable biogas into electricity and heat

  - adequate to power 2,800 homes

Benefits

- Converts waste disposal problem into multiple value streams
- Ultra-clean power facilitates clean air permitting
- Distributed generation enhances power reliability and energy security
- High efficiency drives economics

2.8 MW power plant at Inland Empire municipal water treatment facility Ontario, California

> *"The clean electrical generation process and the reliable 24/7 operating nature of the fuel cell will help us attain the objectives of our strategic energy plan and position us to meet ever more stringent clean air emission requirements," said* ***Terry Catlin, Board President, Inland Empire Utilities Agency.***

19

FuelCell Energy
Ultra-Clean, Efficient, Reliable Power

Integrated Biomass Liquid Fuel Production

Clean CO_2 Increases Biomass Production

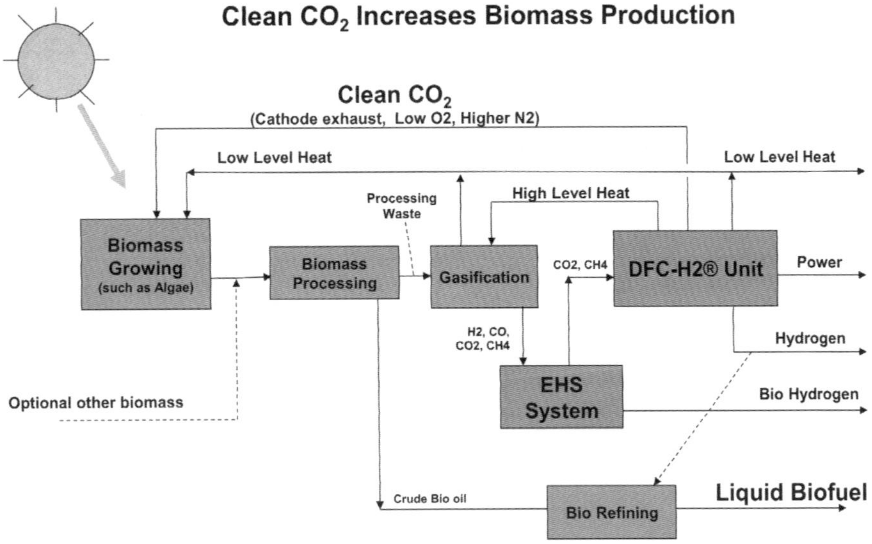

20

FuelCell Energy
Ultra-Clean, Efficient, Reliable Power

Low Carbon On-Site Power from Natural Gas

- 3.8MW ultra low carbon system for on-site power generation from natural gas – larger systems possible
- Captured CO_2 can be sequestered or sold commercially

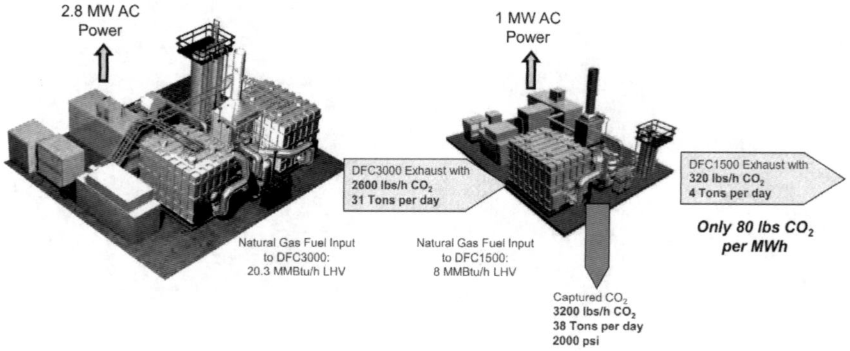

21

Natural Gas Opportunities and Applications for Sustainable Transportation

U.S. DEPARTMENT OF ENERGY | Energy Efficiency & Renewable Energy

Alternative Natural Gas Applications Workshop

Alexandria, VA
October 8, 2014

Reuben Sarkar

U.S. Department of Energy
Deputy Assistant Secretary
Sustainable Transportation

EERE Areas of Focus

Sustainable TRANSPORTATION | Renewable ELECTRICITY GENERATION | Energy Saving HOMES, BUILDINGS, & MANUFACTURING

2

U.S. DEPARTMENT OF ENERGY | Energy Efficiency & Renewable Energy

"All of the Above" for Sustainable Transportation

Sustainable
TRANSPORTATION

- Improving Efficiency
- Diversifying Fuel Sources
- Domestic & Renewable

Hydrogen and Fuel Cells

Vehicles

Bioenergy

National Energy Goals & Climate Action Plan

50% ↓ oil imports
17% ↓ GHG emissions
by 2020

3 U.S. DEPARTMENT OF ENERGY | Energy Efficiency & Renewable Energy

Natural Gas for Sustainable Transportation

- ➢ **Current Status**
- ➢ **Opportunities**
- ➢ **Challenges**
- ➢ **Accomplishments**
- ➢ **Conclusion**

4 U.S. DEPARTMENT OF ENERGY | Energy Efficiency & Renewable Energy

Natural Gas is an Abundant Domestic Source

452

TCF (trillion cubic feet) proven NG reserves*

~2,700

trillion ft³ of total technical and economic potential

Proved Natural Gas (NG) Reserves by State*

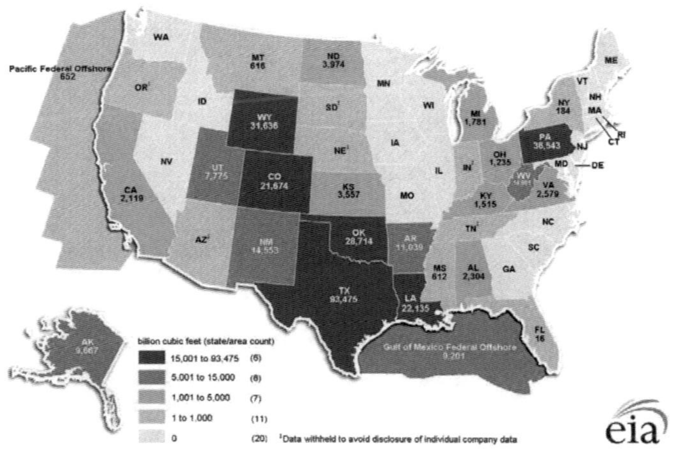

U.S. EIA, Annual Survey of Domestic Oil and Gas Reserves. Released on April 2014.

*323 TCF for wet NG and 129 TCF for shale NG

U.S. DEPARTMENT OF ENERGY Energy Efficiency & Renewable Energy

....with a price advantage over gasoline...

Average Retail Fuel Prices in the U.S.

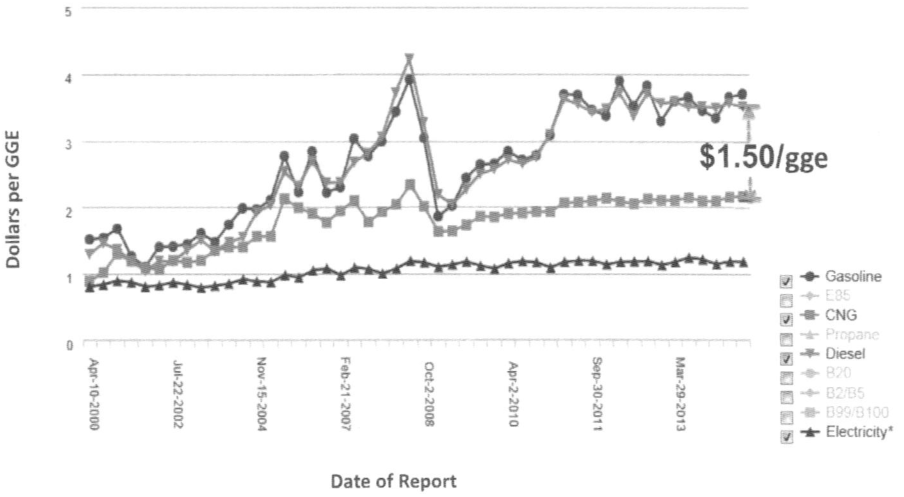

Source: Clean Cities Alternative Fuel Price Reports http://www.afdc.energy.gov/fuels/prices.html

U.S. DEPARTMENT OF ENERGY Energy Efficiency & Renewable Energy

...and with an untapped potential for the transportation sector

U.S. Energy Consumption by Sector

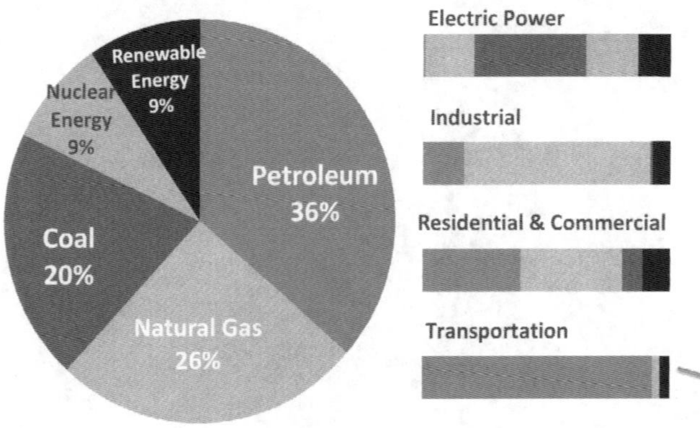

3% of total energy consumed in the transportation sector is **natural gas****

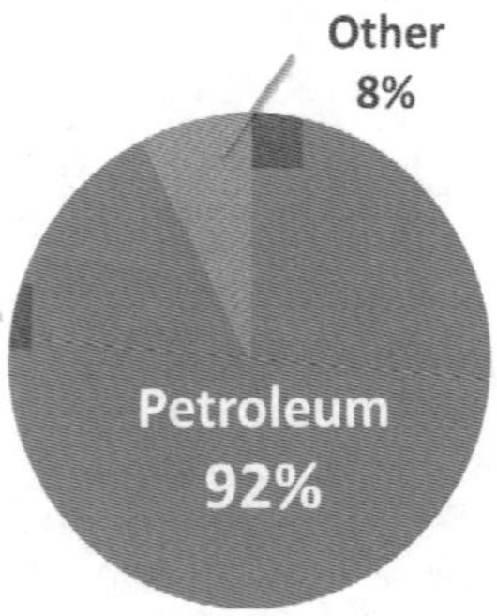

Total U.S. Energy Consumption = **97** Quadrillion Btu/year*

*Energy Information Administration, Annual Energy Review 2011, Table 1.3

**U.S. EIA 2013 Annual Energy Review http://www.eia.gov/totalenergy/data/annual/#consumption

7

U.S. DEPARTMENT OF ENERGY | Energy Efficiency & Renewable Energy

Natural Gas for Sustainable Transportation

- Current Status
- **Opportunities**
- Challenges
- Accomplishments
- Conclusion

8

U.S. DEPARTMENT OF ENERGY | Energy Efficiency & Renewable Energy

Expected Growth and Opportunities in Heavy Duty Segment

Transportation – Fastest Growing Sector for NG Consumption

From 40 Bcf in 2012 to 850 Bcf in 2040

Heavy duty trucks - fastest growing segment

Natural Gas Consumption in the Transportation Sector (quadrillion Btu)

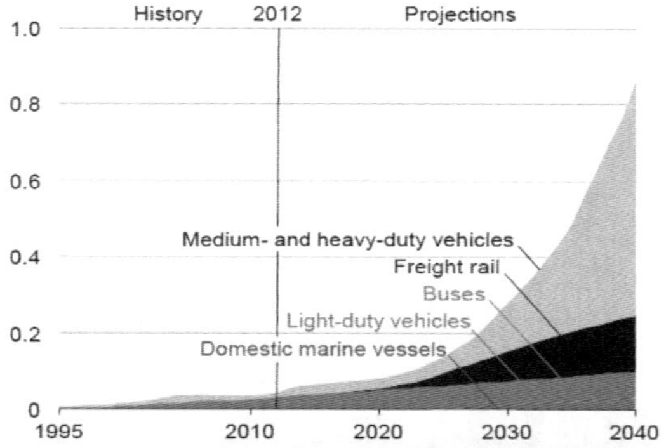

9 Source: EIA, Annual Energy Outlook 2014

U.S. DEPARTMENT OF ENERGY | Energy Efficiency & Renewable Energy

Possible Pathways for Natural Gas Use in Transportation

1. Direct Fuel

Compressed and Liquefied Natural Gas → Fuels → Trucks, Buses, Natural Gas Light Duty Vehicles (NGV)

2. Source to produce alternative fuels

NG → Produces → Electricity → Fuels → BEV

NG → Produces → H_2 → Fuels → H_2 FCV

NG → Produces → Liquid Fuels → Fuels → ICE

10

U.S. DEPARTMENT OF ENERGY | Energy Efficiency & Renewable Energy

Reduced GHG Emissions from Vehicle Portfolio Using NG

Well-to-Wheels Greenhouse Gases Emissions for 2035 Mid-Size Car

(Grams of CO2-equivalent per mile)

NGV

Can result in
~10% ↓
in GHG emissions*

H_2 FCV

With distributed NG is the source of H_2 can result in
>50% ↓
in GHG emissions**

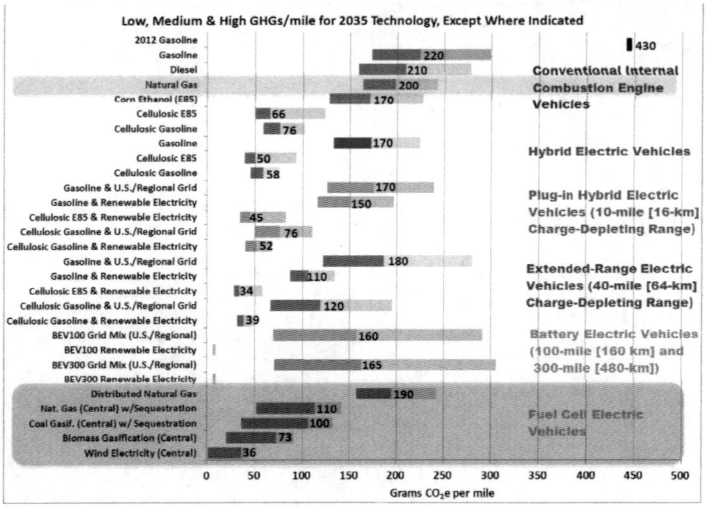

Low/medium/high: sensitivity to uncertainties associated with projected fuel economy of vehicles and selected attributes of fuels pathways, e.g., electricity credit for biofuels, electric generation mix, etc.

Source: http://hydrogen.energy.gov/pdfs/13005_well_to_wheels_ghg_oil_ldvs.pdf

* Compared to 2035 gasoline ICE with NG R&D
** Compared to 2012 gasoline ICE

Largest AFV petroleum reductions come from CNG Vehicles

Clean Cities
U.S. Department of Energy

>50%
petroleum reductions coming from NGVs

~230M
GGE saved with NGVs

Clean Cities Petroleum Savings by AFV Type

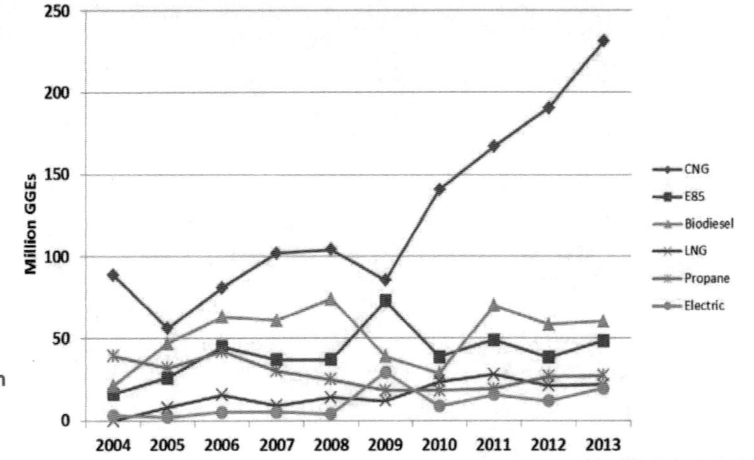

www.afdc.energy.gov/data/

Natural Gas for Sustainable Transportation

- ➢ Current Status
- ➢ Opportunities
- ➢ **Challenges**
- ➢ Accomplishments
- ➢ Conclusion

Infrastructure

~750 CNG and ~65 LNG public stations for NGV

~10 public H_2 stations for FCV

VS.

150,000 Gasoline stations

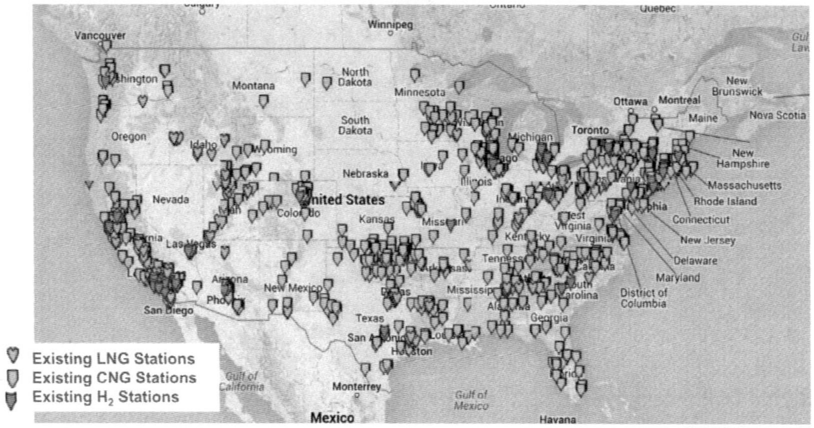

http://www.afdc.energy.gov/fuels/natural_gas_locations.html
http://maps.nrel.gov/transatlas

On-Board Storage

Natural Gas and Gasoline Vehicle Performance Comparison

2014 Honda Civic	Natural Gas	Gasoline
Fuel Capacity (gge)	8.03	13.2
Volume of tank (gal)	~30	~13.2
EPA rated range (miles)	250	410
MSRP	$26,640	$18,390

Price Difference: ~$8,000

- **Space**
- **Driving Range**
- **Cost**

Honda Civic – Natural Gas

Adapted from NPC Future Transportation Fuel Study, NG Subcommittee, 2012

15 U.S. DEPARTMENT OF ENERGY | Energy Efficiency & Renewable Energy

Engine Efficiency of Natural Gas ICEs

- Efficiency penalty of 20% compared to diesel engines
- Efficiency improvements should focus on lean combustion
  - **Ignition Technologies**
  - **Fuel Injection Technologies**
- Near-term target is less than 5% efficiency penalty compared to diesel engines
  - Longer-term is parity

*Overall Efficiency value in the long term

Source: Memorandum for the Secretary, EERE Planned activities in NG as a transportation fuel in FY16

16 U.S. DEPARTMENT OF ENERGY | Energy Efficiency & Renewable Energy

Natural Gas for Sustainable Transportation

- Current Status
- Opportunities
- Challenges
- **Accomplishments**
- Conclusion

17 U.S. DEPARTMENT OF ENERGY Energy Efficiency & Renewable Energy

Clean Cities Awards are Enabling NGVs Deployments

- **$300M Clean Cities ARRA Awards**
  - **20 awards**
  - **36 states**
  - Enabled deployment of **>150** CNG and LNG stations

> 3,600 

- **$11M Clean Cities Alternative Fuel Implementation Awards**
  - **20 awards**
  - **30 states + DC**
  - Helped communities establish action plans, implement sustainable policies, address barriers, and develop incentives
- **$4.5M Current FOA for Alternative Fuel Vehicles**
  - Demonstration and Enhanced Driver Experience Projects
  - Training for First Responders, Public Safety Officials, and Critical Service Providers
  - Emergency Response and Preparedness Operations

18 U.S. DEPARTMENT OF ENERGY Energy Efficiency & Renewable Energy

ARPA-E Projects in NG are Helping Address Unique Challenges

arpa·e
CHANGING WHAT'S POSSIBLE

MOVE Program

Methane Opportunities for Vehicular Energy

The **Advanced Research Projects Agency-Energy (ARPA-E)** advances high-potential, high-impact energy technologies that are too early for private-sector investment.

ARPA-E's MOVE Program is aimed to find cost effective ways to power light duty vehicles with America's abundant natural gas resources.

Objectives

- **Reduced payback for** 

- **Conformable tanks with energy density- CNG**
- **Convenient, low-cost at-home refueling**

Vehicle Technology Forums & User Groups for NGVs

- **Natural Gas Transit and School Bus Users Group**
  - Transit agencies, school bus fleets, and government agencies come together to receive technical assistance and share information about using natural gas
- **Natural Gas Vehicle Technology Forum**
  - Supports development and deployment of commercially competitive natural gas engines, vehicles, and infrastructure.
  - Stakeholders include equipment manufacturers, national laboratories, government agencies, vehicle fleets, and industry groups.

NGVTF
Natural Gas Vehicle Technology Forum

Fuel Cell Cars are Here

FCEVs on display at North American auto shows.

Honda Fuel Cell Electric Vehicle

Toyota Fuel Cell Electric Vehicle

Hyundai's first mass-produced Tucson Fuel Cell SUVs arrive in Southern California May 20, 2014

Lease includes H_2 and maintenance.

21

U.S. DEPARTMENT OF ENERGY | Energy Efficiency & Renewable Energy

H_2 from NG can be Competitive with Gasoline

H_2 from NG using Steam Methane Reforming (SMR)

1. Can make H_2 competitive with gasoline at high volumes
2. Must be produced at < $4/gge by 2020*

Cost of H_2 from Distributed Production by Selected Technologies

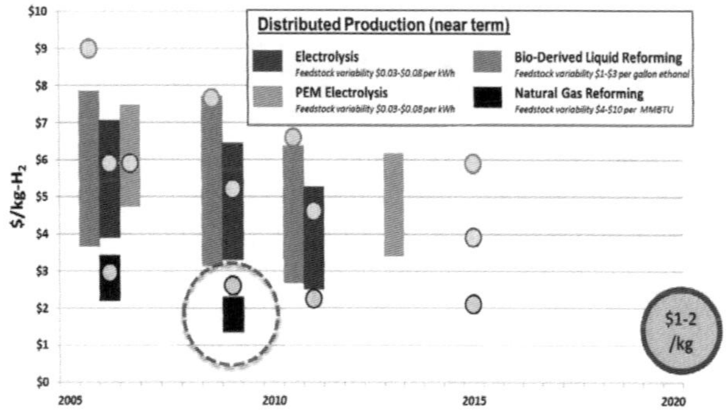

Source: Program Record #10001, www.hydrogen.energy.gov/program_records.html.

*Including delivery and dispensing at the pump

22

U.S. DEPARTMENT OF ENERGY | Energy Efficiency & Renewable Energy

Public-Private Partnership to Overcome H_2 Infrastructure Challenges

H_2USA

Mission: To promote the commercial introduction and widespread adoption of FCEVs across America through creation of a public-private partnership to overcome the hurdle of establishing hydrogen infrastructure.

Natural Gas for Sustainable Transportation

- ➢ Current Status
- ➢ Opportunities
- ➢ Challenges
- ➢ Accomplishments
- ➢ **Conclusion**

Conclusion

- **Abundant domestic NG resources** present an opportunity to reduce our dependence on foreign oil for transportation sector.
- **NG could be a potential bridge** for other alternative fuel vehicles such as FCEVs fueled by H_2 from NG.
- **Synergies between H_2 and NG** offer opportunity to leverage technology so that common challenges are addressed in parallel.

25

U.S. DEPARTMENT OF ENERGY Energy Efficiency & Renewable Energy

Thank you

26

U.S. DEPARTMENT OF ENERGY Energy Efficiency & Renewable Energy

AIChE
The Global Home of Chemical Engineers

CENTER FOR ENERGY INITIATIVES
An AIChE Technological Community

Alternative Natural Gas Applications Workshop:

Creating a Prosperous Demand Market

Transportation Roundtable Discussion

October 9, 2014

NG as a Transportation Fuel

1. **What are the *current uses* for NG in transportation and what are the current estimated sizes of these markets?**
2. **What are *potential areas for expansion* for NG in transportation and what is the future potential of these new markets?**
3. **Who are the *current major players* and leads for NG transportation applications?**
4. **What are the major *technical and regulatory barriers* or roadblocks that prevent growth of NG as a transportation fuel?**
5. **What are the primary *technical gaps* inhibiting wider NG utilization and in what areas?**
6. **What is the magnitude or the *level of impact of the technology gaps*? i.e. How do they compare with other political, social or environmental barriers?**
7. **What *role should the Government play* in addressing potential technology gaps? (i.e. What role can the National Laboratories play in addressing the technical gaps?)**
8. **To *prioritize the efforts*, what are the top three technical gaps which need to be addressed first?**

AIChE
The Global Home of Chemical Engineers

2

1. What are the *current uses* for NG in transportation and what are the current estimated sizes of these markets?

1. **On Road**
   - Class 8 – largest use/growth area
     - *Heavy duty flee - >0.1%x3.9MMG/day*
     - *Refuse haulers 40-50% of new sales CNG*
     - *Transit 30-40% of new sales CNG*
   - Light duty (Class 2 to Class 7)
     - *Commuters / Consumers – no current market (5+ to 10+ year possibility)*
       - Bifuel vehicles
       - Dependent on competitive alternative fuels (as well as traditional gas)
     - *Fleet vehicles / Taxis*

Uses for Natural Gas:
- CNG
- LNG
- GTL
- Biomass to Liquids
  - Diesel / Gasoline alternatives
- Methanol
- Alcohol fuels
- BiFuel
- Dual Fuel

AIChE
The Global Home of Chemical Engineers

3

1. What are the *current uses* for NG in transportation and what are the current estimated sizes of these markets?

2. **Off Road**
   - Marine (regional vs. national markets)
     - *Regional example: Tug boats in Pittsburg*
   - Rail
   - Mining operations
3. **Refueling stations**
   - CNG at 3600 psi
     - *95% of natural gas vehicle market*
     - *Standard refueling is 3 – 5 gge/min*
   - LNG is $0.50 to $0.70 higher than CNG
   - Limitations on NG pipeline based on season/time of year/location
     - *On site location or Public access locations*

Uses for Natural Gas:
- CNG
- LNG
- GTL
- Biomass to Liquids
  - Diesel / Gasoline alternatives
- Methanol
- Alcohol fuels
- BiFuel
- Dual Fuel

AIChE
The Global Home of Chemical Engineers

4

2. What are *potential areas for expansion* for NG in transportation and what is the future potential of these new markets?

1. **Heavy Duty CNG highest near term 10% HD truck sales by 2020**
   - Return to base
     - *Refuse trucks / Buses / Municipal Fleets*
     - *Regional hauling*
   - Coast-to-Coast
     - *Hub-and-spoke transportation*
     - *Network of refueling stations necessary*
     - *Shippers versus carriers… shippers driving market*
   - Retrofitting/Conversions of high-mileage vehicles
   - Dual Fuel (range / refueling guarantee)

Largest areas for market growth:
- Heavy duty vehicles / large trucks / fleets
- Off Road market / Marine / Rail /etc.
- Component manufacturers

Wish list for future growth:
- Weight of pressure vessel
  - Penalty of up to 1000 – 1500 lb
  - Can increase vehicle Class, requiring CDL
- Efficiency of engine
  - (CNG is 6 mpg vs. diesel at 8 mpg)
- Off-board infrastructure improvements
  - 10 gge/min or better fill rate
  - Temperature issues during refueling

AIChE
The Global Home of Chemical Engineers

5

2. What are *potential areas for expansion* for NG in transportation and what is the future potential of these new markets?

2. **Heavy Duty LNG declining sales**
   - Coast-to-Coast
     - *Similar market penetration points*
     - *Higher energy density / longer range than CNG, but higher cost*
3. **Mid Duty**
4. **Off Road**
   1. Marine
      - *Fishing industry (single-day round trip applications)*
      - *Emission controls/reductions drive in this market*
   2. Rail
   3. Privately owned LDV

Largest areas for market growth:
- Heavy duty vehicles / large trucks / fleets
- Off Road market / Marine / Rail /etc.
- Component manufacturers

Wish list for future growth:
- Weight of pressure vessel
  - Penalty of up to 1000 – 1500 lb
  - Can increase vehicle Class, requiring CDL
- Efficiency of engine
  - (CNG is 6 mpg vs. diesel at 8 mpg)
- Off-board infrastructure improvements
  - 10 gge/min or better fill rate
  - Temperature issues during refueling

AIChE
The Global Home of Chemical Engineers

6

3. Who are the *current major players* and leads for NG transportation applications?

1. **Consumers**
   - Heavy Duty - Swift, UPS, Dillon, Saddle Creek, Frito Lay, P&G, J.B. Hunt
   - Refuse - Waste management, Republic Services, Waste Connections
   - Transit- NYC, Boston, Washington, Richmond, Cleveland and many more
2. **OEMs**
   - Ford
   - Chrysler
   - GM
   - Volvo
   - Honda
   - Freightliner
   - Peterbilt
   - Kenworth
   - Toyota
3. **Second Tier MFG. & QVM**
   - CumminsWestport
   - Rousch
   - Caterpillar
4. **Suppliers of NG (LNG, CNG, etc.)**
   - Shell
   - Exxon
   - Chevron
   - NG Utilities
   - BP
5. **NG Investors**
   - Trillium CNG

AIChE
The Global Home of Chemical Engineers

7

4. What are the major *technical and regulatory barriers* or roadblocks that prevent growth of NG as a transportation fuel?

1. **Regulatory**
   - Inconsistent interpretations of codes and standard (NFPA 52) requirements
     - *Evaluation of codes and standards impact (planned or unforeseen)*
     - *Level of involvement – city, county, state, or federal*
     - *Inconsistent measure of energy usage / billing*
       - Desire an absolute standard of comparison: dge vs. gge vs. BTU vs. kg vs. ???
       - Consumers are accustomed to volume measurements (gallon, liter, etc.)
   - Regulatory uncertainty from state and federal government
     - *Policies based on factors other than science / in opposition to science*
     - *<u>Consistent</u> tax incentives / credits could accelerate innovation (5+ year window)*
   - Utility tariffs are inconsistent, and at times irrational
   - Variability of technology incentives / investment
     - *Levels of research – fundamental research through commercialization*
     - *Standardize targets and technologies*
       - Competition of components / options / features
   - Carbon Taxes / Café Standards / Penalties / Incentives
     - *Markets most affected will have greatest change / growth / reduction*

4. What are the major *technical and regulatory barriers* or roadblocks that prevent growth of NG as a transportation fuel?

2. **Commercial**

- High incremental cost of the truck
- NG storage: mass, volume, package, and cost
  - *Pressure vessel / tank*
  - *Accompanying technologies*
- Fast full fill (> 10 gge/min fill rate)
- Fuel Availability (Network of refueling stations, hub-and-spoke, home refueling, etc.)
- Supply of vehicles, engines & parts
  - *Dual Fuel technologies*
- Engine efficiency and cost versus gasoline / diesel / other alternative fuels
- Competitiveness versus gasoline / diesel / other alternative fuels
  - *LNG is not market efficient except in certain applications*
    - $0.50 to $0.70 more per gge than CNG, but higher energy capacity / range
  - *Advances in CNG or ANG can overtake LNG*

A

The (

5. What are the primary *technical gaps* inhibiting wider NG utilization and in what areas?

1. CNG engine tuning and truck specifications – improved MPG (10-15% less than diesel)
2. Adsorbed Natural Gas – where weight is not an issue, lower pressure and conformable tanks – lowers infrastructure and vehicle costs
3. CNG Pressure Vessel / Tank Cost, Mass, Volume, and Package
   - Safety certification cost
   - Reduction in Mass/Volume of storage while maintaining / increasing storage energy / capacity
     - *Maintain safety and reliability*
     - *Cost of composite material*
     - *Manufacturing processes of composite pressure vessels (Type 3 and Type 4)*
   - Reduction in Cost
   - Change in packaging (ANG, conformable tanks, etc.) without cost increase
4. Heat of compression – will improve fill efficiency, reduce truck weight and upfront cost
   - Cooling technologies (at refueling station, within vessel, outside of vessel)
   - Compression technologies – cost, energy use, life cycle
5. Robust SIDI conversion to NG

AIChE
The Global Home of Chemical Engineers

10

5. What are the primary *technical gaps* inhibiting wider NG utilization and in what areas?

6. Refueling Station economics
   - Compressor technology cost
   - Refueling speed must be maintained / increased (> 10 gge/min)
7. Cost / Technical Barriers to alternative NG fuels
   - Ethanol, Methonal, GTL, and other liquid fuels

AIChE
The Global Home of Chemical Engineers

11

6. What is the magnitude or the *level of impact of the technology gaps*? i.e. How do they compare with other political, social or environmental barriers?

1. Small in comparison. As the market grows, innovators will step in and improve products and performance.

AIChE
The Global Home of Chemical Engineers

12

7. What *role should the Government play* in addressing potential technology gaps?
(i.e. What role can the National Laboratories play in addressing the technical gaps?)

1. Consistent, long term and visionary federal and state policy. Regulatory and policy certainty is key to investment.
2. From a National Labs perspective - The market needs good data. Work on getting and distributing that so the market can make good decisions and move forward.
3. Short and Long term R&D Needs
   - NG Storage
     - *Decrease in composite tank manufacturing costs and composite material cost*
     - *Conformable tanks*
     - *Adsorbent materials / low pressure technologies*
     - *Compression technologies*
   - Infrastructure
     - *Compatibility with complimentary fuels at existing and new refueling stations*
     - *Pipeline quality – regional, seasonal, and economic variability*
       - Varying levels of ethane within NG
     - *Home refueling Technology*

AIChE
The Global Home of Chemical Engineers

13

7. What *role should the Government play* in addressing potential technology gaps?
(i.e. What role can the National Laboratories play in addressing the technical gaps?)

3. Short and Long term R&D Needs (continue)
   - Engine Technology
     - *Efficiency improvements in dedicated NG engines*
     - *High temperature material components and lubricants*
     - *Ignition technologies*
     - *Lean burn technologies*
     - *Methane oxidation catalyst*
     - *RCCI*
     - *Adaptive control*
     - *Dual and Tri Fuel Engines*

AIChE
The Global Home of Chemical Engineers

14

8. To *prioritize the efforts*, what are the top three technical gaps which need to be addressed first?

1. Fuel economy – diesel engine technology has been evaluated for 60 years. Natural gas engines for 2 years. The evolution of the natural gas engine needs to continue, improve both SI and develop dual fuel technologies.
2. Tank storage weight versus heat of compression valuation and technology road mapping.
3. High Hp NG engines for on- and off-road markets

- *Market failure is lack of private investor/manufacturer willing to put forth money to address this market*
- *Accelerated catalyst aging for certification*
- *DICI engine conversion technologies*
- *Fuel supply logistics*

AIChE
The Global Home of Chemical Engineers

15

Back-up Slides

Alternative Natural Gas Applications Workshop
Transportation

2010 US Nat. Gas Consumption

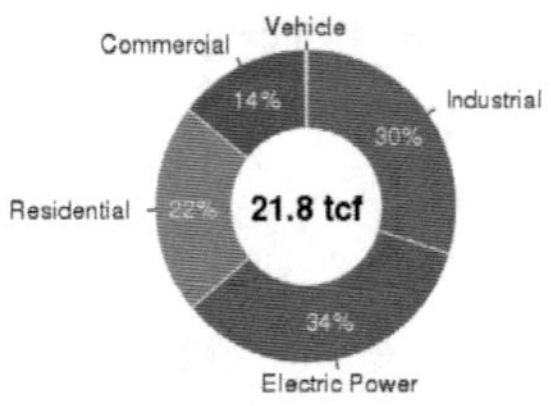

Data: EIA Graphic: mazamascience.com

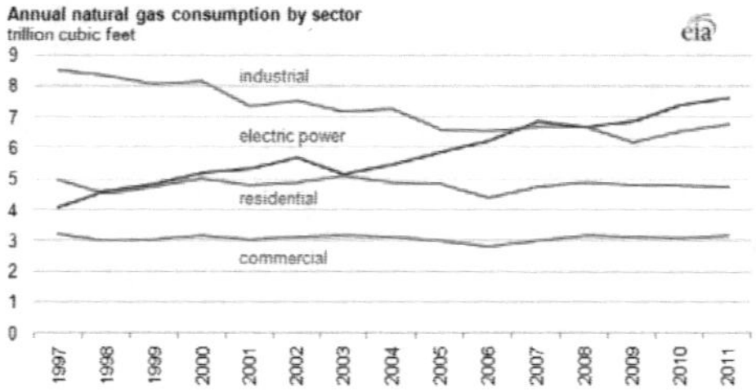

How is NG used today?

- The largest share goes to electrical power and it is the fastest growing market today

CENTER FOR ENERGY INITIATIVES
An AIChE Technological Community

Alternative Natural Gas Applications Workshop
Transportation

What is the most efficient use of NG?

- Producing electricity from NG is perhaps the least efficient.
- What about Vehicles?

Natural Gas delivered to your home instead of being converted to electricity conserves three times as much of the original natural energy!

NATURAL GAS = **90%** Energy System Efficiency

100,000 Btus - Less 10,000 Btus transmitted through pipelines = 90,000 Btus Delivered to Your Home

ELECTRICITY = **27%** Energy System Efficiency

100,000 Btus - Less 73,000 Btus converting electricity and moving it through transmission lines = 27,000 Btus Delivered to Your Home

CENTER FOR ENERGY INITIATIVES
An AIChE Technological Community

Alternative Natural Gas Applications Workshop

Transportation

AIChE
The Global Home of Chemical Engineers

Which is more efficient EV vs. CNG Vehicles

- Producing electricity from NG is perhaps the least efficient.

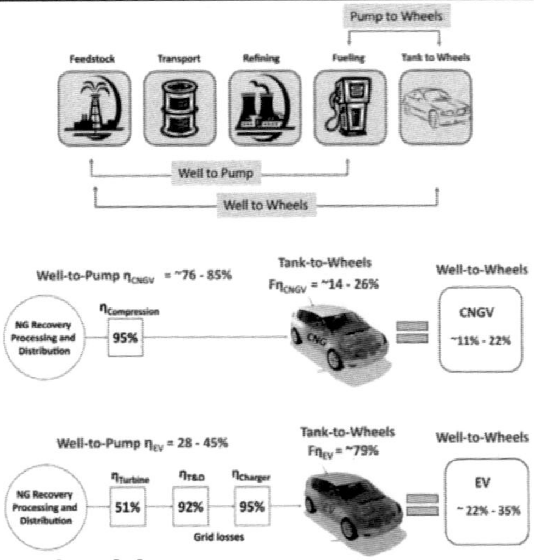

Source: Gas2.org

What about hybrids, hydrogen...?

What about GHG emissions?

CENTER FOR ENERGY INITIATIVES
An AIChE Technological Community

Alternative Natural Gas Applications Workshop

Transportation

AIChE
The Global Home of Chemical Engineers

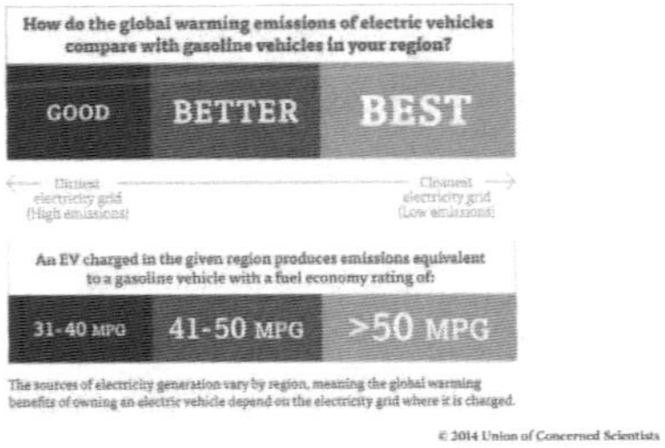

GHG emissions for EV in equivalent MPG for gasoline vehicles.

What about CNG?

What about LNG vs. Diesel?

CENTER FOR ENERGY INITIATIVES
An AIChE Technological Community

Chemical Synthesis

Discussion Centered On

- NG Future Impact, Value Proposition, Roles, Metrics, Technical Barriers, Defining Success, National Benefit, Public Perception, ...

Some Discussion Results

- **Clear Policy** - Develop concrete policies for supporting RD&D to enable technology options for NG in chemical industry.
- **Sustainable Funding Policy** - Fund for the long term to allow development and sustainability of programs that meet the goals.
- **Focus**
  - Short: Increased efficiency, separations
  - Mid-long: Displacement of Oil-based polymers
- **Maturation** – To a level where assessment can be accomplished.
- **Research Agenda** - Prioritizing the Federal research agenda to have the greatest impact, and look at industry as customers.

AIChE
The Global Home of Chemical Engineers

CENTER FOR ENERGY INITIATIVES
An AIChE Technological Community

Alternative Natural Gas Applications Workshop:

Creating a Prosperous Demand Market

Stationary Power Gen Roundtable Discussion

October 9, 2014

NG as a Stationary Power Gen Fuel

1. **What are the *current uses* for NG in Stationary Power and what are the current estimated sizes of these markets?**
2. **What are *potential areas for expansion* for NG in stationary power gen and what is the future potential of these new markets?**
3. **Who are the *current major players* and leads for NG stationary power gen applications?**
4. **What are the major *technical and regulatory barriers* or roadblocks that prevent growth of NG as a stationary power gen fuel?**
5. **What are the primary *technical gaps* inhibiting wider NG utilization and in what areas?**
6. **What is the magnitude or the *level of impact of the technology gaps*? i.e. How do they compare with other political, social or environmental barriers?**
7. **What *role should the Government play* in addressing potential technology gaps? (i.e. What role can the National Laboratories play in addressing the technical gaps?)**
8. **To *prioritize the efforts*, what are the top three technical gaps which need to be addressed first?**

1. What are *potential areas for expansion* for NG in stationary energy generation and what is the future potential of these new markets?

For NG Turbines

- Improved efficiency for combined cycles as pointed out in the Tonko Bill
- The major variable in this is whether US government support of increased R&D funding materializes. If it doesn't happen, the share of the NGCC market filled by US-based manufacturers is projected to drop to 42% from 84% today by 2037 (Exhibit ES-1) and 36,000 fewer US jobs by 2035 (Exhibit ES-2).

For Fuel Cells

- Reliable and resilient distributed power generation is an emerging market where NG has the demonstrated advantage for use as clean fuel. Technologies such as SOFC can efficiently utilize NG in CHP and electricity generation in scales ranging from small commercial to large distributed generation (hundreds of kWe to MWe).

Cross-cutting Areas

- CHP
- Polygen, where products in addition to electric power are produced.
  - Oxygen + Power
  - CO_2 for Enhanced oil recovery + power
  - H_2 + power + CO_2

AIChE
The Global Home of Chemical Engineers

3

2. Who are the *current major players* and leads for NG in stationary energy generation applications?

For NG Turbines:

- The major players in NGCCs are US-based manufacturers GE, Siemens and Alstom; and Japanese supplier Mitsubishi Heavy Industries (MHI). For smaller gas turbines (< 5-10 MW_{th}) which are often used in simple cycle or combined heat & power applications the major players are GE, Rolls-Royce, Solar Turbines (part of Caterpillar) and PW Power Systems (part of MHI).

For Fuel Cells:

- Bloom Energy, FuelCell Energy, LG Fuel Cell and GE Fuel Cells

CCS:

- Mitsubishi and Cansolv

Polygen suppliers

- Air Products; Praxair

AIChE
The Global Home of Chemical Engineers

4

3. What are the major *regulatory, or other barriers* or roadblocks that prevent growth of NG in stationary energy generation?

GHG Mitigation is an Uncertain factor -

- Current proposed regulations exempt NG generation, but if GHG mitigation goals are to be reached, eventually regulations will require mitigation from NG-fired sources. The interim period where NG is not subject to GHG regulations provides an ideal time for R&D targeted at reducing costs for carbon capture utilization and storage so that when regulations are put into place, industry will be ready to comply. The problem will only cost more if we wait to address; will not just go away.

Difficulty in connecting CHP to the Utility Grid

- High charges for back-up power; low prices for produced power

AIChE
The Global Home of Chemical Engineers

5

4. What are the major *technical barriers* or roadblocks that prevent growth of NG in stationary energy generation?

For NG Turbines:

- Major technical areas are combustion, heat transfer, materials – thermal barrier coatings). It's very much the overall efficiency goal that includes the 3 barriers.

For Fuel Cells:

- Technical gaps include (1) deployment of clean and efficient energy conversion system such as SOFC, (b) development and deployment of OTM, (c) reclamation of water used during NG extraction. Technology development areas include functional ceramics and membranes, catalysts, robust filtration processes.

For CCS:

- DOE is actively funding coal focused CCS R&D but should consider funding gas focused CCS R&D to overcome the differences in flue gas content. Demonstration would alleviate some of the risks associated with the technology.

AIChE
The Global Home of Chemical Engineers

6

5. What science or engineering research and development programs should be expanded to address these gaps?

For NG Turbines:

- The Tonko Bill would provide sufficient funding to enable US-based NGCCs of 64% to go on line by 2022 and 67% by 2027. It was introduced in November 2013 in the House. Bill will be reintroduced with next congress. $50M/yr for 7 years ($350M total). Does not fund a demonstration plan as could be too expensive for government input. This would support major component testing.

For Fuel Cells:

- Viable solution to extending the lifetime of fuel cells and reducing the materials and manufacturing costs

Cross-cutting R&D:

- Across the board integrated basic science programs to understand thermal barrier coatings and high temperature materials (these technologies are complimentary in nature so an integrated research is required)

For CCS:

- An R&D program to develop capture technologies specific to NG-fired systems is needed. This program should be focused on accelerating the development of technologies tailored to the unique characteristics of NG-derived flue gases such as lower CO_2 concentrations as well as lower contaminant concentrations.

AIChE
The Global Home of Chemical Engineers

7

6. What is the magnitude or the *level of impact of the technology gaps*? i.e., How do they compare with other political, social or environmental barriers?

For Gas Turbines

- the impact of the technology gaps are significant. As pointed out in Question #1, the gaps equate to the loss of 42 percent loss of market share of US-based manufacturers and 36,000 US jobs by 2035.

For Fuel Cells:

- Commercialization of the technology is currently in sync with the national focus on enhancing the manufacturing ability and increasing the employment

Energy-water nexus – need to show how NG utilization for stationary power gen impacts water utilization. The need to extend life-cycle analyses for NG utilization technologies is critical.

AIChE
The Global Home of Chemical Engineers

8

7. What *role should the Government (i.e., federal and state) play* in addressing potential technology or policy gaps? (i.e., What role can the National Laboratories play in addressing the technical gaps?)

For NG Turbines:
The National Energy Technology Laboratory (NETL) would be the major government player in the proposed Tonko Bill which would address this issue for Gas Turbines. Presumably NETL would manage the program, as they did for the quite successful Advanced Gas Turbine Systems Research (AGTSR) Program which ran from about 1992 to 2000 to demonstrate major components.

For Fuel Cells:

- Developing programs to attack the key technical problems associated with implementing the fuel cell technology in next generation systems; fund the R&D that is beyond near term vision (i.e., 5 years horizon) along with demonstration projects efficient and cost effective NG utilization for stationary applications, for example in polygen, to gain market confidence and acceptance and De-carbonized fuel demonstration projects

Public–private partnership under the guidance of federal/ state government ; university research and industrial participation at the R&D and demonstration level should be examined. National labs will play key role in developing big science and engineered systems development of accelerating the deployment.

8. To *prioritize the efforts*, what are the top three technical gaps which need to be addressed first?

Develop and demonstrate cost effective GHG reduction strategies related to NG Stationary Power Gen (all technologies gas turbines, fuel cells and CCS) with parallel targets for reduction over time. A goal could be to demonstrate a 20% improvement CAPEX and OPEX over current state of the art which will enable further investment.

AUTHOR INDEX

Alger, Monty 10
Carolan, Michael F. 48
Day, William 49
Huang, Kevin 32
Jones, Nick 1
Munson, Ron 53
O'Sullivan, Francis 23
Patel, Pinakin 58
Sarkar, Reuben 69

AIChE
120 Wall Street, FL 23
New York, NY 10005-4020

ISBN 978-1-5108-1765-4

9 781510 817654

CURRAN ASSOCIATES INC.
proceedings.com

Fuels and Petrochemicals Division 2014

Core Programming Area at the 2014 AIChE Annual Meeting

Atlanta, Georgia, USA
16-21 November 2014

AIChE
The Global Home of Chemical Engineers